Studies in Systems, Decision and Control

Volume 124

Series editor

Janusz Kacprzyk, Polish Academy of Sciences, Warsaw, Poland
e-mail: kacprzyk@ibspan.waw.pl

The series "Studies in Systems, Decision and Control" (SSDC) covers both new developments and advances, as well as the state of the art, in the various areas of broadly perceived systems, decision making and control- quickly, up to date and with a high quality. The intent is to cover the theory, applications, and perspectives on the state of the art and future developments relevant to systems, decision making, control, complex processes and related areas, as embedded in the fields of engineering, computer science, physics, economics, social and life sciences, as well as the paradigms and methodologies behind them. The series contains monographs, textbooks, lecture notes and edited volumes in systems, decision making and control spanning the areas of Cyber-Physical Systems, Autonomous Systems, Sensor Networks, Control Systems, Energy Systems, Automotive Systems, Biological Systems, Vehicular Networking and Connected Vehicles, Aerospace Systems, Automation, Manufacturing, Smart Grids, Nonlinear Systems, Power Systems, Robotics, Social Systems, Economic Systems and other. Of particular value to both the contributors and the readership are the short publication timeframe and the world-wide distribution and exposure which enable both a wide and rapid dissemination of research output.

More information about this series at http://www.springer.com/series/13304

Abdellah Benzaouia · Fouad Mesquine
Mohamed Benhayoun

Saturated Control of Linear Systems

Springer

Abdellah Benzaouia
Department of Physics, Faculty of Science Semlalia
University Cadi Ayyad
Marrakech
Morocco

Mohamed Benhayoun
Department of Physics, Faculty of Science Semlalia
University Cadi Ayyad
Marrakech
Morocco

Fouad Mesquine
Department of Physics, Faculty of Science Semlalia
University Cadi Ayyad
Marrakech
Morocco

ISSN 2198-4182 ISSN 2198-4190 (electronic)
Studies in Systems, Decision and Control
ISBN 978-3-319-88145-4 ISBN 978-3-319-65990-9 (eBook)
https://doi.org/10.1007/978-3-319-65990-9

Printed on acid-free paper

This Springer imprint is published by Springer Nature
The registered company is Springer International Publishing AG
The registered company address is: Gewerbestrasse 11, 6330 Cham, Switzerland

To Driss, Tijani and Ouafa
To the memory of Omar

Abdellah Benzaouia

To the Memory of My Mother and My Father
To Sanaa, Marwa, Hiba and Mohamed Yassine
To My Sisters and My Brother
To all My Family

Fouad Mesquine

To the Memory of My Mother and My Father
To My Wife Naima
To My Pediatrician, My Dentist and My Engineer, who will recognize themselves

Mohamed Benhayoun

Preface

When beginning the project of this book, we were wondering why the community would be interested in reading it knowing that constrained control or limited input is a field of abundant and various result papers and books. But, as the project was evolving and the chapters contents making precise, we guessed and hoped that the book would be read for the facts we try to clarify as follows: First, the book has as a leading line the problem of constraints on the inputs which is widely studied these last years but presents also different related problems in control of such systems. Further, the book attempts to gather all the recent results about constrained inputs that are becoming essential for practical reasons more than theoretical ones. Moreover, we are interested in another kind of constraints, namely rate or increment limitations that are becoming very challenging in control applications. Furthermore, we tried in this book to present all eventual cases that may face an engineer or a researcher in an application of control for constrained input systems. It can be quoted that in the presence of limitations, one may be asked to deal with uncertainties, non-measurable states, singularities, delay, two dimension systems, etc. Hence, we will present the robustness of the obtained feedback controllers, the use of observers or output feedback. We will also handle the problem of singular systems and delay systems when the inputs are limited within given sets. Two-dimensional systems, commonly known as 2D systems, will also be studied.

The aim of this book is to give an overview of all the works developed in our team related to constrained control over last two decades. Major differences about this book and works treating the problem of constrained inputs are as follows: First, the constraints on increment or rate of control are introduced. The increment or rate constraints are not nested as it is studied in similar works but both constraints, on the input and its increment or rate, are in parallel. Second, positive invariance-based results are given leading to algebraic conditions that are easy to check but with a trials and error procedure. As presented, linear programming may be used to overcome this withdrawing. Another way for avoiding this problem is the introduction of LMI conditions. In fact, in this case with the given conditions, the stabilization problems become feasibility problems easily checked with the available software's like MATLAB. For both cases of handling constrained control and as it is the vein of all constrained control methods, the enlargement of initial condition set is obtained. Third and not lastly as a second part, convex writing of the closed-loop system having constraints on the input, introduced in the recent literature, will be also used leading to LMI conditions to design stabilizing controllers for such systems.

Marrakech, Morocco
May 2017

Abdellah Benzaouia
Fouad Mesquine
Mohamed Benhayoun

Acknowledgements

We want to thank all our colleagues who have indirectly been working with us to realize this book. Our great thanks go to all our colleagues who worked with us on the subject of saturated control systems, in occurrence, we cite Profs. F. Tadeo, A. Hmamed, D. Mehdi, and M. Darouach. We would like to thank our previous Ph.D. students for their indirect help. This book would not be realizable without the sabbatical year accorded by the Faculty of Sciences Semlalia of University Cadi Ayyad to M. Benhayoun.

Abdellah Benzaouia, Fouad Mesquine, Mohamed Benhayoun
Marrakech, Morocco
May 2017

Contents

List of Figures

Abbreviations and Notations

Abbreviations

AS	Assumption
CCRPOF	Constrained Continuous-time Regulator Problem via Observed State Feedback
CL	Closed-Loop
CTC	Continuous-Time Case
DC	Direct Current
DTC	Discrete-Time Case
iff	If and Only if
IR	Impulse Response
LCRP	Linear Constrained Regulator Problem
LMI	Linear Matrix Inequality
LPC	Linear Programming in Continuous-time Systems Case
LPD	Linear Programming in Discrete-time Systems Case
LS	Linear System
LTI	Linear Time Invariant
MIMO	Multi-Input Multi-Output
OP	Open-Loop
PD	Proportional Derivative
RLCRP	Robust Linear Constrained Regulator Problem
SISO	Single Input Single Output
SLS	Singular Linear System
SS	Singular System
w.r.t.	With Respect to
WWTP	Wastewater Treatment Plants

Notations

- For $x \in \mathbb{R}^n$ a vector:
  - $|x|$ the vector of absolute value of x.
  - x^T its transpose.
  - $x^+ = (x^+)_j$ and $x^- = (x^-)_j$ for $j = 1, \ldots, n$, where $(x^+)_j = \sup(x_j, 0)$ and $(x^-)_j = \sup(-x_j, 0)$.
  - $\delta x(.)$ denotes its derivative with respect to time in the continuous-time case or $x(t+1)$ in the discrete-time case.
- For two vectors $x, y \in \mathbb{R}^n$, $x \leq y$ if $x_i \leq y_i$, $i = 1, \ldots, n$.
- For a matrix $H = (h)_{ij}$ $i, j = 1, \ldots, m$, the tilde transforms are defined by:

$$\widetilde{H}_d = \begin{bmatrix} H^+ & H^- \\ H^- & H^+ \end{bmatrix},$$

where $H^+ = (h^+)_{ij}$, $H^- = (h^-)_{ij}$, $i, j = 1, \ldots, m$ and

$$\widetilde{H_c} = \begin{bmatrix} H_1 & H_2 \\ H_2 & H_1 \end{bmatrix}$$

with:

$$\begin{cases} (H_1)_{ii} = h_{ii} \\ (H_1)_{ij} = h_{ij}^+ \text{ for } i \neq j \end{cases} \text{ and } \begin{cases} (H_2)_{ii} = 0 \\ (H_2)_{ij} = h_{ij}^- \text{ for } i \neq j \end{cases}$$

- H_j, $j = 1, \ldots, m$, denotes the jth row of matrix H.
- $\sigma(H)$ denotes the spectrum of matrix H.
- $\lambda_{min}(H)$, $\lambda_{max}(H)$ are the minimum and the maximum eigenvalue of H, respectively.
- $\mathbb{I}$ is the identity matrix of appropriate size.
- D denotes the stability domain for eigenvalues, that is, the left half plane in the continuous-time case (CTC) or the unit disk in the discrete-time case (DTC).
- $int\mathbb{R}^n_+$ is the interior of $\mathbb{R}^n_+$
- $\mathbb{R}^{n*} = \mathbb{R}^n - \{0\}$
- The set $\mathcal{M}_H$ denotes the set of non-negative matrices for the DTC and the set of matrices with non-negative off-diagonal elements for the CTC (i.e., matrices H such that $h_{ij} \geq 0\ \forall i \neq j$).
- For a matrix $H \in \mathcal{M}_H$ we use $\mathcal{H}$ to note: $H - \mathbb{I}$ in the DTC, and H in the CTC.
- For two matrices A and B of $\mathbb{R}^{n \times m}$, $A \leq B$ if $A(i, j) \leq B(i, j)$, $i = 1, \ldots, n$, $j = 1, \ldots, m$
- For a matrix P of $\mathbb{R}^{n \times n}$ $P \succ 0$, $(P \succeq 0)$ means P positive definite (positive semidefinite, respectively).
- A_{cl} denotes the closed-loop system matrix, while A_i^{cl} represents the closed-loop system matrix with a subscript.

- $\odot$ denotes the null matrix of appropriate dimension.
- Scalar η, vector e and $\mathcal{I}$ denote: $\eta = 2^m$, $e = [1, \ldots, 1]^T \in \mathbb{R}^m$, $\mathcal{I} = \{1, \ldots, \eta\}$
- For $X \in \mathbb{R}^{n \times n}$: $X^{sym} = X + X^T$
- $*$ stands for the transpose of the non-diagonal element Y in the LMI,

$$\begin{bmatrix} X & Y \\ * & Z \end{bmatrix} \succ 0$$

- For two matrices $A \in \mathbb{R}^{n \times m}$ and $B \in \mathbb{R}^{p \times q}$, $A \otimes B$ denotes the Kronecker product:

$$\begin{bmatrix} a_{11}B & \ldots & \ldots & a_{1m}B \\ a_{21}B & a_{22}B & \ldots & a_{2m}B \\ \vdots & \vdots & \vdots & \vdots \\ a_{n1}B & \ldots & \ldots & a_{nm}B \end{bmatrix}$$

- The notation $\triangleq$ is used for defined equality.
- $co\{\cdot\}$ stands for convex hull of $\{\cdot\}$.
- For a function:

$$\begin{aligned} \Phi : [-\tau, 0] &\rightarrow \mathbb{R}^n \\ t &\rightarrow \Phi(t) \end{aligned}$$

we note:

$$\|\Phi(t)\|_c = \sup_{-\tau \leq t \leq 0} \|\Phi(t)\|$$

where the norm $\| \;\|$ stands for the Euclidean norm or the induced matrix norm.
- $\mathcal{C}_{n,\tau}$ is the Banach space of continuous vector functions mapping $[-\tau, \; 0]$ into $\mathbb{R}^n$ with the topology of uniform convergence.
- $x_t \in \mathcal{C}_{n,\tau}$ denotes the restriction of $x(t)$ to the interval $[t - \tau, \; t]$ translated to $[-\tau, \; 0]$.

Introduction and Book Preview

In the field of dynamical control, and in the general case, stabilizing controllers synthesis procedure may be depicted as follows:

- First of all, a model of the system is chosen.
- Goals and requirements of the control system are defined as stability first then secondly performances like steady-state errors, time rising, time response, attenuation levels for a class of signals, and so on…
- Adequate methods are followed for synthesis.
- Simulations are performed to test the obtained control system.
- The controller, as designed, is used for the real system.

For the previous steps of synthesis and during functioning, some problems may occur like:

- The model used for the synthesis is different from the real plant. Two reasons can cause such case like the model is only an approximation of the real system and represent it in a limited zone for the used variables or some parameters may vary when the system is on function.
- The output of the controller is too large for the input of the actuator.
- The variation of the controller output exceeds the ability of the actuator to follow such changes.
- The state used for the model is not measurable and/or no device available or at suitable costs.
- The states exceed the values used for obtaining the linear model, and hence, the linear behavior is no longer valid.
- Some of the states may be related to the others as linear combination leading to singularities in the functioning of the system.

Hence, usually, real or physical plants, as quoted, are subjected to constrained variables. The most frequent constraints are of saturation type: limitations on the magnitude of certain variables. Most practical control problems are dominated by control constraints. Actuator nonlinearities of saturation type are inherent to practically all systems. In fact, and as examples of actuators, one can cite the valves that can operate only between fully open and fully closed positions, voltage and current of servomotors that are limited in given ranges. Further, in several applications, the system to be controlled works in the neighborhood of some operating condition where a linearized model can provide a good approximation, but severe constraints on actuator activity are present. As an afterthought in the design of the controller, it can compromise the control goals. Furthermore, if the control contains an integrator, control saturation leads to the popular phenomenon called integrator windup and its consequences. Ignoring such constraints can be detrimental to the stability and performances of the control system. Effectively, such systems may exhibit some unexpected performance and can even become unstable if the saturation is not taken into account in the system design steps. To avoid such problems, it becomes necessary to take into account actuator saturation at the controller design phase. Consequently, the class of systems with saturation has obtained great interest during the last decades. Many different methods can be used, ranging from general nonlinear control design to saturating linear control. Other approaches deal with robust global stabilization of a class of linear systems with input saturation via gain scheduling or nonlinear feedback [119], and input delay [120–122]. It is worth noticing here that the problem of constrained control of linear systems can be seen as the control problem of nonlinear systems where the saturation is treated as a locally sector-bounded nonlinearity [57, 64]. In these works and references therein, absolute stability analysis tools such as circle and Popov criteria are used. Many approaches that deal with linear feedback stabilization of linear systems subject to input saturation may be quoted as:

a. The optimal control theory where the limitations on the control are taken into account as constraints included in the optimization process [50].
b. The low and high gain-based approach presented by Lin and Saberi [72], where a semiglobal stabilization is sought and is limited to null controllable systems with bounded controls [107], that is, stabilizable systems with all the open-loop poles in the closed left half plane.
c. The ℓ_1 optimization concept [34]. This technique calculates, with simple methods, controllers for different kinds of constraints (limitations in magnitude, slope, overshoot, undershoot, etc) [35, 91, 92, 109, 110].
d. The positive invariance approach where no criterion needs to be minimized. Further, stability assumption on the open-loop system is not necessary. The local nature of this approach can be regarded as semiglobal one for domains of polyhedral type in comparison with the second approach.

e. A convex writing of the constrained system as a linear combination of some linear systems allowing saturation to take effect while guaranteeing asymptotic stability [62].

The main feature of the approach (d) is to confine the state of the system to a set where saturating control does not occur. Since the first paper of Gutman and Hagander [55], conditions of positive invariance of polyhedral sets have been investigated, for instance, by Benzaouia and Burgat [9], Bitsoris [22, 116], and Chegancas [30] for the discrete-time case and by Benzaouia and Hmamed [10], Blanchini [23, 25], and Vassilaki and Bitsoris [116] for the continuous-time case. The application of such conditions to the state feedback control has also been considered, and different kinds of controllers design were proposed by Benzaouia and Burgat [9], Blanchini [23], Burgat and al. [28], Mesquine [79, 82], and Tan and Gilbert [53]. Further, the use of this approach was also extended to the output feedback by Castelan and Tarbouriech [30] and Mesquine and Benzaouia [80]. Besides, a complete characterization of the maximal contractively invariant ellipsoids of linear systems under saturated linear feedback has been presented [70], while the problem with nested saturation has been studied in [112]. Having as final goal, the application of this simple method in an industrial environment, some authors have also studied the robustness of regulators based on positive invariance concept. Works presented in this book will focus on these different aspects, as raised above, during different type of controllers synthesis while taking into account asymmetric constraints on the control magnitude. It may be detailed as follows:

- Asymmetric constraints on the increment or rate of the control;
- Robustness design for controllers with both constraints on the control and parameter uncertainties;
- Output feedback with constraints using or not observers;
- Singular systems;
- Delay systems;
- Two-dimensional systems with delays

.

The book will be interested, especially in two main approaches which have been developed in the literature:

- The first is the so-called positive invariance approach, which is based on the design of controllers which work inside a region of linear behavior where saturation does not occur (see [8, 10, 24] and the references therein). The stabilizing regulator gain F obtained with this approach is usually a solution to the nonlinear algebraic equation $FA + FBF = HF$, where matrix H satisfies some main condition of positive invariance.
- The second approach allows saturation to take effect while guaranteeing asymptotic stability (see [60, 61] and the references therein). This approach, allowing the control to be saturated, leads to a bounded region of stability which, although can be easily obtained by the resolution of a set of LMIs, yet is

ellipsoidal and symmetric. Besides, necessary and sufficient conditions for invariance of convex sets have been studied in [46], while [75] presents a N step set invariance approach to analyze the stability of discrete-time saturated systems.

The main challenge in these two approaches is to obtain a large enough domain of initial states which ensures asymptotic stability of the system despite the presence of saturation [1, 3, 12, 53, 60]. Piecewise quadratic Lyapunov functions approach to estimate the attraction domain for systems with dead-zones or saturation has been presented in [38, 76]. Researchers are still interested in this topic by developing new ideas to deal with [71].

As quoted above, the first problem that this book will focus on is the rate or increment constraints. In fact, apart from saturation constraints, the book deals also with different type of constraints, which were introduced while considering practical applications: incremental or rate constraints on the control variable. As it has been pointed out [2], incremental input saturation is a serious challenge in many automatic control applications, for example, in flight control [42]. In particular, it is known [21] that they can induce a considerable destabilizing effect due to phase-lag. The importance of incremental or rate constraints comes from the fact that, for some processes, the rate of variables change is limited within given bounds. These limits can arise from physical constraints that, if exceeded, could damage the process. Lin [73] showed that all dynamical linear asymptotically null controllable bounded input systems are semiglobally stabilizable through linear feedback in the presence of both constraints. A method to stabilize a particular plant in the presence of constraints on both input magnitude and increments was considered by Trygve et al. [114]. Other approaches have been presented, for example, [45, 63, 74, and 115]. Most of the cited works consider only constraints of symmetric nature. However, the asymmetric character of the actuator rate or increment constraints is very important in practical situations since these constraints are inherently asymmetric: The speed of the actuator is usually different when increasing or decreasing, because the source of the movement is different (in a valve, a spring/air pressure).

The positive invariance approach was selected, for the first time, in [85] and [86] to deal with this problem, where it gives simple methods to calculate constant state feedback controllers with asymmetric constraints and disturbances. However, a different method using the technique developed in [60] has been proposed to take into account nested saturations on the magnitude of the control and its successive derivatives [5]. This problem is different from that solved in the present work where the two saturations are considered in parallel.

As precised before, the book will also focus on the same problem of saturation on the control and asymmetric constraints on its increment or rate and in the context of LMIs presented firstly in [16]. It is based on the previous results of [15, 60, 61, 85, and 86] where the constraints are symmetric.

On the other hand, it is well known that, during the last years, much progress has been accomplished in the regulation methods to take into account model uncertainties. These uncertainties are generally induced by the difference between the real behavior of the system in functioning and the behavior of the plant model used to design the controller. Several techniques have been proposed in the literature, grouped in the theory of the robust control, for example, [43, 48, 65, 118] and the references therein. In addition, during the last decades, combining constraints and uncertainties is of increasing interest to researchers. In fact, the adopted models containing constraints of these systems are often subject to uncertainties which find their origin in the modeling and measurement errors or on the computation approximations. Hence, it is necessary for the control theory to hold into account at the same time constraints and uncertainties. In other words, it is necessary to find robust stabilizing control laws that respect constraints. There are several approaches proposed in the literature to solve this problem but here the positive invariance approach is of interest. Indeed, this approach was extended to take into account uncertainties, and hence, the robustness of such controllers is studied in [11, 23, 83, 84]. Considering the importance of linear programming in matrix algebra and efficient algorithms to solve such problems in the literature ([34] for example), the idea was to translate the results found within the framework of positive invariance under algorithms of linear programming [7, 88, 116].

For the same goal, necessary and sufficient conditions of positive invariance are re-formulated such that linear programming algorithms can be used to find robust constrained regulators for both continuous-time and discrete-time linear systems.

The stability of singular or descriptor linear systems with input saturation and asymmetric constraints is studied in this book. Singular systems have been of great interest in the control literature since they can model many systems in electrical circuits networks, robotic, and economics [37]. However, to the best of our knowledge, few works were dedicated to the study of singular system with constraints on the control (see [51, 104] and the references therein). In these works, necessary and sufficient conditions of positive invariance were given for singular system with state constraint by using the Weierstrass transformation of the original system on an equivalent reduced-order system. Currently, although much effort has been made in the exploration of special properties for singular systems, almost the studies are confined to the generalization of classical system theory. Some problems of observers and synthesis of stabilizing controllers for singular linear systems can be cited in this category [13–90], [51, 101, 104] and the references therein. This is of interest in situations where the derivative can be measured and one wishes to avoid high gain state feedback. In fact, the use of PD controllers has a long history in industrial practice, as it is well known, where derivative controls are employed to provide anticipatory action for overshoot reduction in the responses. Many works have discussed the design of state and derivative feedback controllers for linear systems (see [33, 44, 89] and the references therein). In [36], PD feedback is used to accomplish the objective of shifting all controllable open-loop finite and dynamic infinite modes of descriptor systems to desired finite points. Thus, necessary and sufficient conditions of positive invariance for singular system with constraints on the control and its rate by using a PD controller are presented. As a particular case, results for PD controllers for linear systems with constraints on the control and its rate are obtained.

When the state is not available for the control, the output feedback or the use of observers becomes necessary. Hence, reduced-order observer is introduced in the case of singular linear systems with input saturation and asymmetric constraints. Some problems of observers and synthesis of stabilizing controllers for singular linear systems can be cited in this category [31–90, 101, 104] and the references therein. The problem of constructing an observer for the singular system is an active area for research since two decades: Works like [36, 37, 39, 40, 78, 103] and the references therein may be cited. An observer that further respects the constraints on the control, obtained only with the estimation of the state, will be constructed. The earlier work on regular systems [81] is extended to singular systems.

On the other hand, the second possibility when the state is not available is the use of an output feedback. Despite its apparent simplicity, the problem is still open. A number of numerical procedures have been proposed for solving it since the work of Kimura [66]. A survey was given by [108], and recent progress has been made for the related problem of pole placement; see [49, 52] and the references therein. However, less works were proposed for linear systems with actuator saturations. In [56], a dynamic output feedback is considered, while in [13] and [30], the positive invariance approach is used.

Static output feedback problem for linear systems subject to actuator saturation will be also in interest thereafter. This work extends the results of [29, 61] where a state feedback is used to the case of output feedback. A different proof of the main result of [61] is obtained by using this technique. The synthesis of the controller by static output feedback is also proposed by means of LMIs for linear systems subject to actuator saturations. The proposed technique is completely different from all the previous works cited before on the same subject. The obtained region of invariance and contractivity is generally less conservative.

An attempt to solve the stabilization problem for asymmetrically constrained systems but in the framework of LMIs is also presented. It is worth recalling that the positive invariance approach may handle non-symmetrical constraints but the obtained algebraic conditions cannot be written under LMI form. It is well known that only works using constraints of symmetric nature as in [17, 55, 60, 61] can be expressed under LMI form. However, the asymmetric character of the actuator constraints, which is in concern in this book, is very important in practical situations since these constraints are inherently asymmetric. Many attempts were developed to emphasize LMIs and problems with asymmetric saturations but without great success as in [14, 15]. Hence, the regulator problem for continuous-time linear system with asymmetric constraints on the control in terms of an LMI problem is addressed. Results of [55], easily written under LMIs but restricted to symmetric constraints, are extended to systems with asymmetric constraints and formulated under LMI form for the first time [20].

Nowadays, time delay in dynamical systems is well known as a source of performance degradation and even instability. This issue, i.e., control of time delay systems, has attracted the effort of automatic control community and continue attracting a lot of researcher work developing stability and stabilizability conditions for time delay systems [26, 69, 95]. Combining the delay problem with the problem of constrained control for one system is an attempt to approach a real system where both saturating actuators and time delay may be present. Hence, some results about control of linear systems with both time delay and saturating actuators can be found in the literature. Not exhaustively [32, 58, 96, 113] may be cited. These works are mainly based on measure matrix, complex Lyapunov equations, or still Razmukhin's approach and the set of initial condition where the stability of the closed-loop system can be guaranteed is not given. In [99, 100, 111], the problem is studied using a Lyapunov–Krasovskii technique firstly introduced in [67]. Writing the saturating system as a convex combination of some linear systems is the cornerstone of this approach.

For delay systems, this technique is also used in [99]. Hence, uniform boundedness in the presence of an additive external perturbation is studied. In this book, the same problem is under study, using the approach of these last works. Hence, two delay stability and stabilizability conditions are obtained.

In the last two decades, the 2D system theory has been paid considerable attention by many researchers. The 2D linear models were introduced in the seventies [47, 54] and have found many applications in digital data filtering, image processing [105], modeling of partial differential equations [77], etc. Saturated continuous 2D systems with multi-delays described by Roesser model will be addressed in this book. The stabilization of this kind of systems has been extensively studied in the literature for 1D (see [58, 59, 97] and the references therein). This problem has already been studied for 2D systems by considering delay independent and dependent stability and stabilization conditions [68] and [102]. However, all the studies on 2D delay systems are only available for discrete systems, except authors in [18], where the 2D continuous systems with delay are taken without saturation. To the best of our knowledge, no works on saturated 2D continuous systems with delay exist before [18]. The objective will be to design stabilizing state feedback controllers for this class of systems. To this end, quadratic Lyapunov functions are used. In this context, sufficient conditions of stabilizability under LMI form are presented.

Some case studies will also be in interest in the sequel. Practical plants in the context of the previously presented results are studied.

The first application is devoted to a pH-process control in a tank. Robust controller is deduced and tested in a real laboratory process.

The second application concerns the modeling and control of activated sludge process. This plant is recognized as the most common and major unit process for reduction of organic waste and has become a subject of great interest. Researchers [4, 41, 93, 94, 117] have investigated different control strategies for the monitoring of such processes. The development of effective control strategies on this kind of wastewater treatment plants (WWTP) is hampered by the inherent nonlinearities, the time-varying dynamics, and the lack of suitable instrumentation.

In fact and during the last decades, many investigations have been focused on the control of the nitrogen and dissolved oxygen in an activated sludge reactor within a WWTP with different strategies. One may quote predictive control, optimal control, and adaptive control [6, 98]. Note here that constraints on the control are note handled, and further, all required measurements are assumed available. Apart from this, one may also cite works about the same topic but limited to estimation [27] and not the control. Furthermore, works combining estimation to control for monitoring such processes can also be found [106]. However, constraints are not taken into account during the design steps. Therefore, this application may be thought as a generalization where constraints, estimation, and control are considered using the positive invariance concept together at the design stage. Then, the objective is to apply positive invariance concept techniques to a WWTP [19, 87]. The obtained linearized model combines the problems of non-availability of the state to measure with the limitations of some variables. The control is achieved by an observer-based controller that can take into account constraints on the control and on the error. The obtained linear model is worked out to meet all design required conditions. The efficiency of the process monitoring is shown via simulations with the real plant.

This book is composed of 11 chapters. Chapter 1 presents some preliminary results used in the developments of the others. The robust constrained state and control regulator problem is considered in Chap. 2. Necessary and sufficient conditions of positive invariance are established. A linear programming approach is presented in order to construct, for an uncertain constrained linear systems, a stabilizing linear state feedback control. The control law transfers asymptotically to the origin any initial state belonging to a given set, while constraints in the control vectors are respected.

The regulator problem for linear uncertain systems having state and control constraints is also considered. Necessary and sufficient conditions of positive invariance of polyhedral domains are extended to the case of continuous-time uncertain systems. Robust constrained regulators are then derived.

Chapter 3 is divided into two parts. The first one concerns the problem of constraints on the control together with constraints on the control rate or increment. Linear systems in state space form, in both the continuous-time and discrete-time domains, are considered. Necessary and sufficient conditions are derived for autonomous linear systems with constrained state increment or rate (for the continuous-time case), such that the system evolves respecting rate constraints. A pole assignment technique is then used to solve the inverse problem, giving stabilizing state feedback controllers that respect non-symmetrical constraints on both control and its increment or rate. An illustrative example shows the application of the method on the double integrator problem.

The second part completes the first one by adding a bounded disturbance to the system dynamics. This last part is devoted to the control of linear systems with constrained control and rate or increment with additive bounded disturbances. Hence, the obtained control law is within the admissible intervals for both the control magnitude and its rate or increment and is robust against additive bounded disturbances.

Chapter 4 deals with the problem of singular continuous-time systems with constraints on the control by using a state feedback controller. The problem of controlling singular continuous-time systems with constraints on the control and its rate, by using state and derivative feedback (proportional-derivative controllers), is solved. The synthesis of the controller is presented using two methods: an exact method based on non-symmetric Riccati equations and an approximate method based on the solution of an algebraic equation. An example illustrates the feasibility of the proposed approach. As a particular case, the extension to non-singular systems is also presented.

Chapter 5 deals with the problem of observer-based control design for both regular and singular continuous-time systems with constraints on the control. Necessary and sufficient conditions for the existence of such controllers are obtained. The synthesis is presented using the solution of equation $XA + XBX = HX$ in the regular case and $XA + XBXE = HXE$ in the singular one.

Chapter 6 solves the problem of designing stabilizing regulators for linear systems subject to control saturation and asymmetric constraints on its increment or rate, using reduced dimension LMIs developed on a reduced-order state space. Compared with previous approaches, the proposed technique is valid for asymmetric constraints on the increment or rate of the control, while the computing time is improved by solving reduced dimension LMIs.

Chapter 7 presents sufficient conditions of asymptotic stability of discrete-time linear systems subject to actuator saturation with an output feedback law. The obtained results are given under LMI formulation. A new proof is presented to obtain previous conditions of asymptotic stability. A numerical example is used to illustrate this technique by using a linear optimization problem subject to LMI constraints. The continuous-time case, with output feedback, is also presented as a straightforward extension. Example is given to illustrate the obtained conditions.

Chapter 8 deals with the regulator problem for both discrete-time and continuous-time linear system with asymmetric saturation on the control. The main contribution of this chapter is to extend the available results, in the LMI form, for symmetrical saturation to the case of unsymmetrical saturation. A new transformation for constrained input linear problem control is presented to deal with the asymmetry of the constraints. Hence, LMI formalism is obtained for the first time for asymmetrical saturation. An example is presented, in each case, to illustrate the obtained results.

In Chap. 9, two results about saturating delay systems are presented. Both are based on the convex writing of the system as convex combination of linear delay systems. The first condition is delay independent and introduces the partitioning of the delay interval to obtain a less conservative condition. The second is delay dependent and it extends an improved existing condition to the case of saturating systems.

Chapter 10 deals with the problem of stabilizability with saturated control of 2D continuous systems with multi-delays. State feedback control is used. Sufficient conditions for asymptotic stability are presented. The synthesis of the required controllers is given under LMI form. An illustrative example is treated.

The aim of Chap. 11 is to present case study where some of the presented results are applied on real processes. First, a pH control in a tank is treated and robust controller is derived using positive invariance techniques.

The second application is devoted to the control of a nonlinear biological nitrogen removal process. Design steps of an observer-based control scheme applied to the linearized model of a phenomenological model of the process are illustrated. The estimation algorithm is combined with the control technique to monitor the process. The goal of the control is the removal or at least the reduction of organic waste. The control law is based on positive invariance concept that had shown efficiency in handling control constraints. The efficiency of both the control and the estimation is demonstrated via computer simulations.

The book is ended by some concluding remarks.

References

1. M. Ait Rami, H. Ayyad and F. Mesquine, Enlarging ellipsoidal invariant sets for constrained linear systems. Int. J. Innovative Comput. Inform. Control **3**(5), 2007
2. D. Angeli, A. Casavola, and E. Mosca, Predictive PI-control of linear plants under positional and incremental input saturation. Automatica **36**, 1505–1516 (2000)
3. A. Baddou, F. Tadeo, A. Benzaouia, On improving the convergence rate of linear constrained control continuous-time systems with a state observer. IEEE Trans. Circ. Syst. I fundam. Theory Appl (2008)
4. G. Bastin and D. Dochain, On line estimation and adaptive control of bio-reactors (Elsevier, Amsterdam, 1990)
5. A. Bateman and Z. Lin, An analysis and design method for linear systems under nested saturation. Syst. Control Lett. **48**, 41–52 (2003)
6. C.A.C. Belchiora, R.A. Araujo and J.A. Landeck, Dissolved oxygen control of the activated sludge waste water treatment process using stable adaptive fuzzy control. Comput. Chem. Eng. **37**, 152–162 (2012)
7. L. Benvenuti and L. Farina, Constrained control for uncertain discrete-time linear systems. Int. J. Robust and Non-linear Control **8**, 555–565 (1998)
8. A. Benzaouia and C. Burgat, Regulator problem for linear discrete-time systems with non-symmetrical constrained control. Int. J. Control **48**, 2441–2451 (1988)
9. A. Benzaouia and C. Burgat, Existence of nonsymmetrical Lyapunov functions for linear systems. Int. J. Syst. Sci. **20**, 597–560 (1989)
10. A. Benzaouia and A. Hmamed, Regulator problem for linear continuous systems with non-symmetrical constrained control. IEEE Trans. Autom. Control **38**(10), 1556–1560 (1993)
11. A. Benzaouia and F. Mesquine, Regulator problem for uncertain linear discrete-time systems with constrained control. Int. J. Robust and Nonlinear Control **4**, 387–395 (1994)
12. A. Benzaouia and A.A. Baddou, Piecewise linear constrained control for continuous-time systems. IEEE Trans. Autom. Control **44**(7), 1477–1481 (1999)
13. A. Benzaouia and D. Mehdi, The output feedback saturated controller design for linear systems. Med. Conf., Portugal (2002)
14. A. Benzaouia, M.A. Rami and S. El Faiz, Stabilization of linear systems with saturation: A Sylvester equation approach. IMA J. Math. Control Inform. **21**(3), 247–259 (2004)
15. A. Benzaouia and S. El Faiz, The regulator problem for linear systems with constrained control: An LMI approach. IMA J. Math. Control Inform. **23**(3), 335–345 (2006)

16. A. Benzaouia, F. Tadeo and F. Mesquine, The regulator problem for linear systems with saturation on the control and its increments or rate: An LMI approach. IEEE Trans. Circ. Syst. I fundam. Theory Appl. **53**(12), 2681–2691 (2006)
17. A. Benzaouia, F. Mesquine, A. Hmamed and H. Aoufoussi, Stability and control synthesis for discrete-time linear systems subject to actuator saturation by output feedback. Math. Prob. Eng. J. (2006)
18. A. Benzaouia, M. Benhayoun and F. Tadeo, State feedback stabilization of 2D continuous systems with delays. IJICIC **7**(2), 977–988 (2011)
19. A. Benzaouia, Saturated switching systems. LNC **426**, Springer, ISBN-13: 978-1447128991 (2012)
20. A. Benzaouia, M. Benhayoun and F. Mesquine, Stabilization of systems with unsymmetrical saturated control: An LMI approach. Circ. Syst. Sig. Process. **33**(10), 3263–3275 (2014)
21. J.M. Berg, K.D. Hammett, C.A. Schwartz and S.S. Banda, An analysis of the destabilizing effect of daisy chained rate limited actuators. IEEE Tran. Control Syst. Tech. **4**(2), 171–176 (1996)
22. G. Bitsoris, Positively invariant polyhedral sets of discrete time linear systems. I. J. Control **4**, 387–395 (1988)
23. F. Blanchini, Feedback control for linear time-invariant systems with state and control bounds in the presence of disturbances. IEEE Trans. Autom. Control **35**, 1231–1234 (1990)
24. F. Blanchini, Set invariance in control. Automatica **35**, 1747–1767 (1999)
25. F. Blanchini and S. Miani, Set-theoretic methods in control. Birkhäuser (2008)
26. E.K. Boukas and Z. Liu, Deterministic and stochastic time delay systems. Birkhauser, Boston (2002)
27. B. Boulkroune, M. Darouach, M. Zasadzinski, S. Gille and D. Fiorelli, A nonlinear observer design for an activated sludge wastewater treatment process. J. Process Control **19**, 1558–1565 (2009)
28. C. Burgat, S. Tarbouriech and M. Klai, Continuous time saturated feedback regulators theory and design. Int. J. Syst. Sci. **25**, 315–336 (1994)
29. Y.Y. Cao and Z. Lin, Stability analysis of discrete-time systems with actuator saturation by saturation dependent Lyapunov function. Proc. of the 41th IEEE CDC, Las Vegas, USA (2002)
30. B. Castelan and S. Tarbouriech, On positive invariance and output feedback stabilization of input constrained linear systems. ACC **3**, 2740–2744 (1994)
31. M. Chaabane, O. Bachelier, M. Souissi and D. Mehdi, Stability and stabilization of continuous descriptor systems. An LMI approach. Math. Prob. Eng. **2006**, 1–15 (2006)
32. B.S. Chen, S.S. Wang and H.C. Lu, Stabilization of time delay systems containing saturating actuators. Int. J. Control **47**, 867–881 (1988)
33. Chu and M. Malabre, Numerically reliable design for proportional and derivative state feedback decoupling controller. Automatica **38**, 2121–2125 (2002)
34. M.A Dahleh and I.J. Diaz-Bobillo, Control of uncertain systems: a linear programming approach. Englewoods Cliffs, NJ: Prentice Hall (1995)
35. M.A. Dahleh and J.B. Pearson, ℓ_1-Optimal feedback controllers for MIMO discrete-time systems. IEEE Trans. Autom. Control **32**, 314–322 (1987)
36. L. Dai, Observers for discrete singular systems. IEEE Trans. Autom. Control **33**(2), 187–191 (1988)
37. L. Dai, Singular control systems. Lect. Notes in Control Inform. Sci. (Springer Verlag, Berlin, 1989)
38. D. Dai, T. Hu, A.R. Teel and L. Zaccarian, Output feedback design for saturated linear plants using deadzone loops. Automatica **45**(12), 2917–2924 (2009)
39. M. Darouach, Solution to Sylvester equation associated to linear descriptor systems. Syst. Control Lett. **55**(9), 835–838 (2006)
40. M. Darouach, M. Zasadzinski and M. Hayar, Reduced-order observer design for descriptor systems with unknown inputs. IEEE Trans. Autom. Control **41**(7), 1068–1072 (1996)

41. D. Dochain, Design of adaptive controller for nonlinear stirred tank bioreactors: Extension to the MIMO situation. J. Process Control **1**, 41–48 (1991)
42. M.A. Dornheim, Report pinpoints factors leading to YF-22 crash. Aviat. Week Space Tech. **9**, 53–54 (1992)
43. J.C. Doyle, K. Glover, P.P. Khargonekar, and B.A. Francis, State space solution to H_2 and H_∞ control problems. IEEE Trans. Autom. Control **34**(8), 831–847 (1989)
44. M.B. Estrada and M. Malabre, Proportional and derivative state feedback decoupling of linear systems. IEEE Trans. Autom. Control **45**, 730–733 (2000)
45. G. Feng, M. Palaniswami, and Y. Zhu, Stability of rate constrained robust pole placement adaptive control systems. Syst. Control Lett. **18**, 99–107 (1992)
46. M. Fiacchini, C. Prieur and S. Tarbouriech, Necessary and sufficient conditions for invariance of convex sets for discrete-time saturated systems. 52th IEEE CDC, Florence, Italy (2013)
47. E. Fornasini and G. Marchesini, State-space realization theory of two-dimentional filters. IEEE Trans. Autom. Control **21**(4), 484–492 (1976)
48. A. Francis and G. Zames, Design of H_∞ optimal multivariable feedback systems. Proc. CDC, 103–108 (1983)
49. G. Garcia, P. Pradin and F. Zeng, Stabilization of discrete-time linear systems by static output feedback. IEEE Trans. Autom. Control **46**(12), 1954–1958 (2001)
50. G. Gautier and G. Bornard, Commande multivariable en présence de contraintes de type inégalitées. RAIRO Automatique **17**(13), 205–222 (1983)
51. C. Georgiou and N.J. Krikelis, A design approach for constrained regulation in discrete singular systems. Syst. Control Lett. **17**, 297–304 (1991)
52. L. El Ghaoui, F. Oustry and M. Ait Rami, Cone complementarity algorithm for static output-feedback and related problems. IEEE Trans. Control **42**(8), 1171–1176 (1997)
53. E.G. Gilbert and K.T. Tan, Linear systems with state and control constraints: The theory and application of maximal output admissible sets. IEEE Trans. Autom. Control **36**, 1008–1020 (1991)
54. D.D. Givone and R.P. Roesser, Multidimensional linear iterative circuits—General properties. IEEE Trans. Comput. **21**(10), 1067–1073 (1972)
55. P.O. Gutman and P. Hagander, A new design of constrained controllers for linear systems. IEEE Trans. Autom. Control **30**(1), 22–33 (1985)
56. D. Henrion, S. Tarbouriech and G. Garcia. Output feedback robust stabilization of uncertain linear systems with saturating controls: An LMI Approach. IEEE Trans. Autom. Control **44** (11), 509–516 (1999)
57. H. Hindi and S. Boyd, Analysis of linear systems with saturating control using convex optimization. Proc. 37th IEEE CDC, Florida, USA (1998)
58. A. Hmamed, A. Benzaouia and H. Bensalah, Regulator problem for linear continuous-time delay systems with nonsymmetrical constrained control. IEEE Trans. Autom. Control **40**, 1615–1619 (1995)
59. A. Hmamed, Constrained regulation of linear discrete-time systems with time-delay: delay-dependent and delay independent conditions. Int. J. Syst. Sci. **31**, 529–536 (2000)
60. T. Hu and Z. Lin, The equivalence of several set invariance conditions under saturation. Proc. 41th CDC, Las Vegas, USA (2002)
61. T. Hu, Z. Lin and B.M. Chen, Analysis and design for discrete-time linear systems subject to actuator saturation. Syst. Control Lett. **45**, 97–112 (2002)
62. T. Hu and Z. Lin, Control systems with actuator saturation. Control eng., Birkhauser (2001)
63. P. Kapasouris and M. Athans, Control systems with rate and magnitude saturation for neutrally stable open loop systems. Proc. IEEE CDC, Hawaii, USA, 3404–3409 (1990)
64. H. Khalil, Nonlinear systems. Upper Saddle River, NJ Prentice Hall (1996)
65. V.L. Kharitonov, Asymptotic stability of an equilibrium position of a family of systems of linear differential equations. Diffrential nye Uravnenya **14**, 2086–2088 (1978)

66. H. Kimura, Pole assignment by gain output feedback. IEEE Trans. Autom. Control **20**(4), 509–516 (1975)
67. N.N. Krasovskii, Stability of motion. Stanford Univ Press CA (1963)
68. W. Lam, J.K. Galkowski, S. Xu and Z. Lin, Robust stability and stabilization of 2D discrete state-delayed systems. Syst. Control Lett. **51**, 277–291 (2004)
69. J.F. Lafay and J. Conte, Analysis and design methods for delay systems. Proc. 34th IEEE CDC, New Orleans, 2035–2069 (1995)
70. Y. Li and Z. Lin, A complete characterization of the maximal contractively invariant ellipsoids of linear systems under saturated linear feedback. IEEE Trans. Autom. Control **60** (1), 179–185 (2015)
71. Y. Li and Z. Lin, Stability and performance analysis of saturated systems via partitioning of the virtual input space. Automatica **53**, 85–93 (2015)
72. Z. Lin and A. Saberi, Semi global exponential stabilization of linear discrete time systems subject to input saturation via linear feedback. Syst. Control Lett. **24**, 125–132 (1995)
73. Z. Lin, Semi-global stabilization of linear systems with position and rate-limited actuators. Syst. Control Lett. **30**, 1–11 (1997)
74. Z. Lin, M. Pachter, S. Banda, and Y. Shamash, Stabilizing feedback design for linear systems with rate limited actuators. Control of uncertain systems with bounded inputs, Lect. Notes Control Inform. Sci., vol. **227** (Springer Verlag, 1997), pp. 173–186
75. Z. Li, Y. Xi and Z. Lin, An Nth-step set invariance approach to the analysis and design of discrete-time linear systems subject to actuator saturation. Syst. Control Lett. **60**(12), 943–951 (2011)
76. Y. Li and Z. lin, A generalized piecewise quadratic lyapunov function approach to estimating the domain of attraction of a saturated system. IFAC **48**(11), 120–125 (2015)
77. W. Marszalek, Two dimensional state-space discrete models for hyperbolic partial differential equations. Appl. Math. Models **8**, 11–14 (1984)
78. B. Marx, D. Koenig and J. Ragot, Design of observers for Takagi-Sugeno descriptor systems with unknown inputs. IET Control Theory Appl. **1**(5), 1487–1495 (2007)
79. F. Mesquine, Contribution à la commande des systèmes dynamiques discrets avec contraintes sur les entrées par application du concept d'invariance positive. Third cycle Cadi Ayyad University thesis (1992)
80. F. Mesquine and A. Benzaouia, Existence of output feedback for a class of systems with constrained control. First I.C.E.A. **4**, 108–116, Tizi ouzou, Algeria (1992)
81. F. Mesquine and D. Mehdi, Constrained observer based-controller for linear continuous-time systems. Int. J. Syst. Sci. **27**(12), 1361–1367 (1996)
82. F. Mesquine, Contribution à la commande des systèmes à entrées contraintes par les observateurs et une nouvelle méthodologie de placement de pôles robuste. Cadi Ayyad University Ph.D. thesis (1997)
83. F. Mesquine, A. Benlamkadem and F. Tadeo, Robust constrained control for continuous time systems: an application to a PH process. IEEE CCA Conf., Glasgow, Scotland, UK, 391–396 (2002)
84. F. Mesquine, A. Benlemkadem and A. Benzaouia, Robust constrained regulator problem for linear uncertain systems. J. Dyn. Control Syst. **10**(4), 527–544 (2004)
85. F. Mesquine, F. Tadeo and A. Benzaouia, Regulator problem for linear systems with constraints on the control and its increments or rate. Automatica **40**(8), 1378–1395 (2004)
86. F. Mesquine, F. Tadeo and A. Benzaouia, Regulator constrained control and rate problem for linear systems with additive disturbances. Proc. ACC, Boston (2004)
87. F. Mesquine and O. Bakka, Observer based regulator problem for wwtp with constraints on the control. Int. J. Innovative Comput. Inform. Control **9**(2), 2013
88. B.E.A. Milani and A.N. Carvalho, Robust linear regulator design for discrete-time systems under polyhedral constraints. Automatica **31**, 1489–1493 (1995)
89. R. Mukundan and W. Dayawansa, Feedback control of singular systems—proportional and derivative feedback of the state. Int. J. Syst. Sci. **14**, 615–632 (1983)

90. P.C. Muller and M. Hou, On the observer design for descriptor systems. IEEE Trans. Autom. Control **38**(11), 1666–1671 (1993)
91. M. Naib, A. Benzaouia and F. Tadeo, ℓ_1-control using linear programming for systems with asymmetric bounds. Int. J. Control **78**(18), 1459–1465 (2005)
92. M. Naib, F. Tadeo and A. Benzaouia, Control of systems with asymmetric bounds using linear programming: Application to a hydrogen reformer plant. Math. Probl. Eng. J. (2006)
93. F. Nejjari, A. Benhammou, B. Dahhou and G. Roux, Nonlinear multivariable control of a biological wastewater treatment process. 4th Eur. Control Conf., Bruxelles, Belgique (1997)
94. F. Nejjari, E. Dahhou, A. Benhammou and G. Roux, Nonlinear multivariable adaptive control of an activated sludge wastewater treatment process. Int. J Adapt. Control Sig. Process. **13**, 347–365 (1999)
95. S.I. Niculescu, H^∞ memoryless control with an α stability constraint for time delay systems An LMI approach. Proc. 34th IEEE CDC, New Orleans, 1507–1512 (1995)
96. S.I. Niculescu, J.M. Dion and L. Dugard, Robust stabilization for uncertain time delay systems constraining saturating actuators. IEEE Trans. Autom. Control **41**(5), 742–747 (1996)
97. S.I. Niculescu, Delay effects on stability. A robust control approach. Lect. Notes Control Inform. Sci. (Springer Verlag, Heidelberg, 2001)
98. M. O'Brien, J. Mack, B. Lennox, D. Lovett and A. Wall, Model predictive control of an activated sludge process: A case study. Control Eng. Pract. **19**, 54–61 (2011)
99. S. Oucheriah, Synthesis of controllers for time delay systems subject to actuator saturation and disturbance. J. Dyn. Syst. Meas. Control **125**, 244–249 (2003)
100. S. Oucheriah, Robust exponential convergence of a class of linear delayed systems with bounded controllers and disturbances. Automatica **42**, 1863–1867 (2006)
101. P.N. Paraskevopoulos and F.N. Koumboulis, Observes for singular systems. IEEE Trans. Autom. Control **37**(8), 1211–1215 (1992)
102. W. Paszke, J. Lam, K. Galkowski, S. Xu, Z. Lin, E. Rogers and A. Kummert, Delay-dependent Stability of 2-D state-delayed linear systems. IEEE ISCAS, 2813–2816 (2006)
103. A.M. Perdon and M. Anderlucci, An unknown input observer for singular time-delay systems. 14th Mediterr. Conf. Control Autom., Ancona, Italy, 28–30 June 2006
104. N.E. Radhy, A. Benzaouia and H. Boughari, Constrained state regulation of linear continuous-time singular systems. Syst. Anal. Modell. Simul. **42**(5), 2002
105. R. Roesser, A discrete state-space model for linear image processing. IEEE Trans. Automat. Control **20**, 1–10 (1975)
106. A. Stare, N. Hvala and D. Vrecko, Modeling, identification, and validation of models for predictive ammonia control in a wastewater treatment plant-A case study. ISA Trans. **45**(2), 159–174 (2006)
107. H.J. Sussmann, E.D. Sontag and Y. Yang, A general result on the stabilization of linear systems using bounded controls. IEEE Trans. Autom. Control **39**, 2411–2425 (1994)
108. V.L. Syrmos, C. Abdallah and P. Dorato, Static output feedback: A survey. Proc. IEEE CDC, 837–842 (1994)
109. F. Tadeo and M.J. Grimble, Controller design using linear programming for systems with constraints. Part 1: Tutorial Introduction; Part 2: Controller Design; Part 3: Design Examples, IEE Comput. Control Eng. J. **12**, 273–276 (2002), **13**, 49–52, 89–93 (2003)
110. F. Tadeo, A. Holohan and P. Vega, ℓ_1-Optimal control of a pH plant. Comput. Chem. Eng., S459–S466 (1998)
111. S. Tarbouriech and J.M.G. da Silva, Synthesis of controllers for continuous-time delay systems with saturating controls via LMI's. IEEE Trans. Autom. Control **45**(1), 105–111 (2000)
112. S. Tarbouriech, C. Prieur, and J.M. Gomes da Silva, Stability analysis and stabilization of systems presenting nested saturations. IEEE Trans. Autom. Control **51**(8), 1364–1371 (2006)

113. E. Tissir and A. Hmamed, Further results on the stabilization of time delay systems containing saturating actuators. Int. J. Syst. Sci. **23**, 615–622 (1992)
114. L. Trygve, R. Murray and T.Y. Fossen, Stabilization of integrator chains in the presence of magnitude and rate saturations; a gain scheduling approach. Proc. IEEE CDC, San Diego, USA, 4004–4005 (1997)
115. F. Tyan and D.S. Bernstein, Dynamic output feedback compensation for linear systems with independent amplitude and rate saturations. Int. J. Control **67**(1), 89–116 (1997)
116. M. Vassilaki and G. Bitsoris, Constrained regulation of linear continuous-time dynamical systems. Syst. Control Lett. **13**, 247–252 (1989)
117. H. Zhao, S.H. Issacs, H. Soeberg and M. Kummel, Nonlinear optimal control of an allmating activated sludge process in a pilot plant. J. Process Control **4**, 33–43 (1994)
118. A.L. Zenlentsovsky, Non-quadratic Lyapunov functions for robust stability analysis of linear uncertain systems. IEEE Trans. Autom. Control **39**(1), 1994
119. B. Zhou, Z. Lin and G. Duan, Robust global stabilization of linear systems with input saturation via gain scheduling. Int. J. Robust and Nonlinear Control **20**(4), 424–447 (2009)
120. B. Zhou, G. Duan and Z. Lin, Global stabilization of the double integrator system with saturation and delay in the input. IEEE Trans. Circ. Syst. I: Regular Papers **57**(6), 1371–1383 (2010)
121. B. Zhou, Z. Lin and G. Duan, Global and semi-global stabilization of linear systems with multiple delays and saturation in the input. SIAM J. Control Optim. **48**(8), 5294–5332 (2010)
122. B. Zhou, Z.Y. Li and Z. Lin, Stabilization of discrete-time systems with multiple actuator delays and saturation. IEEE Trans. Circ. Syst. I: Regular papers **60**(2), 389–400 (2013)

Chapter 1
Preliminary Results

1.1 Introduction

In this chapter, all the needed material for the development in the sequel of this book is presented. Hence, positive invariance concept is recalled in cases of both continuous-time and discrete-time systems. The positive invariance concept remains the more used during the last two decades ([3, 5, 6, 9, 11], and the references there in). Further conditions guaranteeing it are studied. The pole assignment techniques often used with positive invariance is also presented. Furthermore, state constraints and D-positive invariance are reminded. Singular systems and saturating systems are also in interest in this chapter.

1.2 Constrained Control

1.2.1 Discrete-Time Systems

In this section, we deal with linear discrete-time systems described by:

$$x_{k+1} = Ax_k + Bu_k, \tag{1.1}$$

with $k \in \mathbb{N}$, x is the state vector in $\mathbb{R}^n$ and u is the constrained control satisfying

$$u_k \in \Omega \subset \mathbb{R}^m,\ m \leq n. \tag{1.2}$$

As it generally occurs in practical situations, the set of admissible controls Ω is an asymmetric polyhedral set defined by:

$$\Omega = \left\{u \in \mathbb{R}^m / - u_{min} \leq u \leq u_{max},\ u_{max},\ u_{min} \in \mathbb{R}^m_+\right\}. \tag{1.3}$$

A. Benzaouia et al., *Saturated Control of Linear Systems*, Studies in Systems, Decision and Control 124, https://doi.org/10.1007/978-3-319-65990-9_1

Matrices A and B are constant and satisfy the assumption:

$$(A, B) \text{ is stabilizable.} \tag{1.4}$$

In order to study the problem control design under inequality constraints, we follow the approach adopted by [15]. Let us first consider the unconstrained case where the regulator problem for system (1.1) consists in the design of a feedback law:

$$u_k = F x_k, \text{ with } F \in \mathbb{R}^{m \times n}. \tag{1.5}$$

Applying the control law as defined above, system (1.1) becomes:

$$x_{k+1} = (A + BF) x_k = A_{cl}\, x_k \tag{1.6}$$

Let $x(k, x_o)$ be the motion of system (1.6), at time k, starting at x_o.

Generally speaking, the matrix F is chosen in order to speed up the closed-loop system dynamics with (1.6) asymptotically stable, that is:

$$\rho(A + BF) < \rho(A) \text{ and } \rho(A + BF) < 1\,, \\ rank(F) = m$$

In the constrained case, the feedback law is defined by:

$$u_k = sat(F x_k) = \begin{cases} u_{max} & \text{for } F x_k > u_{max} \\ F x_k & \text{for } F x_k \in \Omega \\ -u_{min} & \text{for } F x_k < -u_{min} \end{cases} \tag{1.7}$$

This feedback law implies two possible models for the system in the closed loop:

(i) the linear model:

$$x_{k+1} = (A + BF) x_k = A_{cl} x_k, \qquad \text{for } F x_k \in \Omega, \tag{1.8}$$

(ii) the nonlinear model:

$$x_{k+1} = A x_k + B sat(F x_k), \qquad \text{for } F x_k \notin \Omega. \tag{1.9}$$

Both representations are obtained in two different regions of the state space.

The approach we deal with in this section consists in proceeding in such a way that the model (1.8) remains valid every time. This is only possible if the state is constrained to evolve in a specified region defined by: $F^{-1}\,\Omega = \mathcal{D}(F, u_{max}, u_{min})$. Where $F^{-1}\,\Omega$ stands for the inverse image of the Ω without the requirement of the invertibility of F. From (1.3), (1.7) and (1.8), the set of admissible states is defined as:

$$\mathcal{D}(F, u_{max}, u_{min}) = \{x \in \mathbb{R}^n / u_{min} \leq F x \leq u_{max}\,;\; u_{max}, u_{min} \in \mathbb{R}^m_+\}. \tag{1.10}$$

Note that this domain is unbounded in the general case when $m < n$.

Clearly, if $x_k \in \mathscr{D}(F, u_{max}, u_{min})$ we may get $x_{k+1} \notin \mathscr{D}(F, u_{max}, u_{min})$.

The fact that the state remains in given sets can be meant in the sense of the following definitions.

Definition 1.1 A subset $\mathscr{D}$ of $\mathbb{R}^n$ is said to be positively invariant with respect to (w.r.t.) the system (1.1) and (1.2) if for every initial state $x_o \in \mathscr{D}$, every admissible sequence:

$$\mathscr{U}_k = \{u_o, u_1, \ldots, u_{k-1}; u_i \in \Omega\},$$

the motion $x(x_o, \mathscr{U}_k, k) \in \mathscr{D}$, for every $k \in \mathbb{N}$.

Definition 1.2 A subset $\mathscr{D}$ of $\mathbb{R}^n$ is said to be:

- contractive w.r.t. the system (1.1) and (1.2), if for every $x_k \in \partial(\tau_k \mathscr{D})$, there exists $\tau_{k+1} > 0$, satisfying $\tau_{k+1} < \tau_k$ such that $x_{k+1} \in \partial(\tau_{k+1}\mathscr{D})$, for every admissible control u_k and $k \in \mathbb{N}$. If $\tau_k < 1$ (resp. $\tau_k > 1$), we say that the set $\mathscr{D}$ is in-contractive (resp. out-contractive) w.r.t. the system.
- attractive for a subset $\mathscr{T}$ of $\mathbb{R}^n$ w.r.t. the system (1.1) and (1.2) if, for every $x_o \in \mathscr{T} \backslash \mathscr{D}$, there exists $k_o \in \mathbb{N}$ such that $x(x_o, \mathscr{U}_k, k) \in \mathscr{D}$, for every $k \geq k_o$, and every admissible sequence $\mathscr{U}_k$.
- globally attractive w.r.t. the system (1.1) and (1.2) if $\mathscr{T} = \mathbb{R}^n$.

Note that the contractivness property defined here is a one step contractivity.

Further, if a Lyapunov function $\mathscr{V}(x)$ is known for the system (1.8), then there always exists a scalar $c \in int\mathbb{R}^+$ such that the set:

$$\mathscr{D}_L = \{x \in \mathbb{R}^n \,/\, \mathscr{V}(x) \leq c\}, \tag{1.11}$$

is a subset of $\mathscr{D}(F, u_{max}, u_{min})$.

In the approach proposed by Gutman and Hagander [15], the necessity of the positive invariance property of domain $\mathscr{D}_L$ w.r.t. the system (1.8) (*i.e.*, $A_{cl}\, \mathscr{D}_L \subset \mathscr{D}_L$), when we are interested in achieving (1.12), requires one to find conditions under which the set $\mathscr{D}(F, u_{max}, u_{min})$ is positively invariant w.r.t. the system (1.8). This will be the main purpose of this section.

Hence, for every $x_k \in \mathscr{D}_L \subset \mathscr{D}(F, u_{max}, u_{min})$, we have $u_k \in \Omega$; consequently, the model (1.8) remains valid. Further, since $\mathscr{V}(x)$ is a Lyapunov function for the system (1.8), then for every $x \in \mathscr{D}_o \subset \mathscr{D}_L$, where $\mathscr{D}_o$ denotes the set of admissible initial states, we obtain $x(k, x_o) \in \mathscr{D}_L, \forall\, k \in \mathbb{N}$ and $x(k, x_o) \rightarrow 0$ as k goes to ∞.

It may be noted, from (1.3), that the set $\mathscr{D}(F, u_{max}, u_{min})$ is generally of a polyhedral asymmetric nature. Thus, the largest domain of admissible initial values of system (1.1) is obtained if:

$$\mathscr{D}_o = \mathscr{D}_L = \mathscr{D}(F, u_{max}, u_{min}). \tag{1.12}$$

The use of a quadratic Lyapunov function only allows one to obtain an ellipsoidal stability domain [15]. The idea of constructing the largest polyhedral stability domain $\mathscr{D}_L \subset \mathscr{D}(F, u_{max}, u_{min})$ was put forward by [13] by using simplicial cones. Its formulation in the symmetrical case was given by [29] who gives necessary and sufficient conditions for the set $\mathscr{D}(F, u_{max}, u_{min})$ with $u_{max} = u_{min} > 0$ to be positively invariant w.r.t. the system.

In the subsequent paragraphs of this section, the necessary and sufficient conditions allowing the design of a regulator for linear discrete-time systems with symmetric and asymmetric constrained control will be presented.

Define first, the null space $\mathscr{K}er(F)$ of F as follows:

$$\mathscr{K}er(F) = \{x \in \mathbb{R}^n \,/\, Fx = 0,\ F \in \mathbb{R}^{m\times n}\}. \tag{1.13}$$

Consider the following state transformation,

$$z_k = F x_k, \qquad F \in \mathbb{R}^{m\times n}, \tag{1.14}$$

then, from (1.8), one can obtain,

$$z_{k+1} = F(A + BF)x_k. \tag{1.15}$$

If a matrix $H \in \mathbb{R}^{m\times m}$ exists such that:

$$FA_{cl} = HF, \tag{1.16}$$

or equivalently,

$$FA + FBF = HF,$$

the n-order dynamical system (1.8) can be transformed to an m-order dynamical system given by:

$$z_{k+1} = Hz_k,\ \ z_k \in \mathbb{R}^m, \tag{1.17}$$

and domain (1.10) becomes:

$$\mathscr{D}(\mathbb{I}, u_{max}, u_{min}) = \{z \in \mathbb{R}^m / -u_{min} \le z \le u_{max},\ \ u_{max}, u_{min} \in \mathbb{R}^m_+\}. \tag{1.18}$$

The positive invariance of domain (1.18) implies necessarily the stability of H. Further, comparing (1.10) with (1.18) leads to

$$x_k \in \mathscr{D}(F, u_{max}, u_{min}) \text{ iff } z_k \in \mathscr{D}(\mathbb{I}, u_{max}, u_{min})\ \forall k \in \mathbb{N}. \tag{1.19}$$

It is obvious, that in this case, the domain (1.10) is positively invariant (resp. positively invariant and contractive) w.r.t. the system (1.8) if and only if domain (1.18) is positively invariant (resp. positively invariant and contractive) w.r.t. the system (1.17).

In this approach, matrices A, B, and F are given while matrix H is obtained as a solution to Eq. (1.16). This approach is known as the direct approach and is based on the state transformation (1.14) leading to Eq. (1.16) and the transformed dynamical system (1.17) with domain (1.18) and property (1.19). Hence, a necessary and sufficient condition for domain $\mathscr{D}(\mathbb{I}, u_{max}, u_{min})$ to be positively invariant w.r.t. the system (1.17) is now presented.

Theorem 1.1 ([7]) *The subset $\mathscr{D}(\mathbb{I}, u_{max}, u_{min})$ of $\mathbb{R}^m$ defined by (1.18) is positively invariant (resp. positively invariant and contractive) w.r.t. the system (1.17) if and only if:*

$$\tilde{H}_d U \leq U \ \ (resp. \ \tilde{H}_d U < U), \tag{1.20}$$

with

$$\tilde{H}_d = \begin{bmatrix} H^+ & H^- \\ H^- & H^+ \end{bmatrix} \tag{1.21}$$

$$U = \begin{bmatrix} u_{max} \\ u_{min} \end{bmatrix} \tag{1.22}$$

and

$$H_{ij}^+ = sup(0, H_{ij}), \ H_{ij}^- = sup(0, -H_{ij}), \text{ for } i, \ j = 1 \dots m. \tag{1.23}$$

Coming back to the initial system, the positive invariance of domain $\mathscr{D}(F, u_{max}, u_{min})$ given by (1.10) w.r.t system (1.17) can be obtained as:

Theorem 1.2 ([5, 7]) *The domain $\mathscr{D}(F, u_{max}, u_{min})$ given by (1.10) is positively invariant w.r.t. system (1.8), if and only if there exist a matrix $H \in R^{m\times m}$, such that*

$$F A + F B F = H F \tag{1.24}$$

$$\tilde{H}_d U \leq U \tag{1.25}$$

with $\tilde{H}_d$ and U are defined by (1.21) and (1.22).

As pointed out above, the use of quadratic Lyapunov function may be restrictive in the sense of the obtained stability domain. Consequently, a non-quadratic Lyapunov function for system (1.17) is presented here. Consider the positive definite non-quadratic function:

$$V(z) = \underset{i}{Max} \ max \left\{ \frac{z_i^+}{w_1^i}, \frac{z_i^-}{w_2^i} \right\} \tag{1.26}$$

$V(z)$ must satisfy the following necessary and sufficient conditions in order to be a Lyapunov function for the system (1.17):

Theorem 1.3 ([5]) *The function V(z) given by (1.26) is a Lyapunov function of system (1.17) if and only if:*

$$\tilde{H}_d \vartheta \le \vartheta$$

where: $\vartheta^t = [w_1^t \ w_2^t]$ *with* $w_1, w_2 \in int\mathbb{R}_+^m$.

It is worth noting that vector ϑ can be chosen as $\vartheta = U$, when constraints are considered.

1.2.2 Continuous-Time Systems

This section recalls the extension of the previously presented results in the discrete-time case to the continuous-time one. Consider the linear continuous-time system represented by the following state-space model:

$$\dot{x}(t) = Ax(t) + Bu(t) \tag{1.27}$$

where $x(t) \in \mathbb{R}^n$ denotes the state vector, $u(t) \in \Omega \subset \mathbb{R}^m$ is the control input restricted (by saturation) to evolve in the polyhedral set given by (1.3).

Definition 1.3 A set S of $\mathbb{R}^n$ is positively invariant w.r.t. motions of the system (1.27), if for every $x_o(t_o) \in S, x(t, x_o, t_o) \in S, \ \forall t > t_o$.

Similarly to the discrete-time case, when using a state feedback control law:

$$u(t) = Fx(t), \ F \in \mathbb{R}^{m \times n} \text{ with } rank(F) = m \tag{1.28}$$

that stabilizes the system

$$\dot{x}(t) = (A + BF)x(t) \tag{1.29}$$

induces the following set of linear behavior in the state space: Positive invariance conditions are recalled as follows:

Theorem 1.4 ([5]) *Domain* $\mathscr{D}(F, u_{max}, u_{min})$ *given by (1.10) is positively invariant w.r.t. system (1.29), if and only if there exist a matrix* $H \in R^{m \times m}$, *such that*

$$FA + FBF = HF \tag{1.30}$$

$$\tilde{H}_c U \le 0 \tag{1.31}$$

with

$$\tilde{H}_c = \begin{bmatrix} H_1 & H_2 \\ H_2 & H_1 \end{bmatrix} \text{ and } U = \begin{bmatrix} u_{\max} \\ u_{\min} \end{bmatrix} \tag{1.32}$$

where

$$H_1 = \begin{cases} h_{ll} \ \ for\ l = k \\ h^+_{lk}\ for\ l \neq k, \ \ where\ h^+_{lk} = sup(h_{lk}, 0) \end{cases}$$
$$H_2 = \begin{cases} 0 \ \ for\ l = k \\ h^-_{lk}\ for\ l \neq k, \ \ where\ h^-_{lk} = sup(-h_{lk}, 0) \end{cases} \quad (1.33)$$

Secondly, to present a non-quadratic Lyapunov function, consider the autonomous linear continuous-time system:

$$\dot{z}(t) = Hz(t), \quad (1.34)$$

and the positive definite non-quadratic function (1.26). $V(z)$ must satisfy the following necessary and sufficient conditions in order to be a Lyapunov function for the system (1.34):

Theorem 1.5 ([5]) *Function V(z) given by (1.26) is a Lyapunov function of system (1.34) if and only if:*

$$\tilde{H}_c\vartheta \leq 0$$

where $\vartheta^t = [w_1^t \ w_2^t]$ *with* $w_1, w_2 \in int\mathbb{R}^m_+$.

It is worth noting that vector ϑ can be chosen as $\vartheta = U$, when constraints are involved.

1.3 Resolution of Equation $XA + XBX = HX$

From the development above, the equation:

$$X A + X B X = H X \quad (1.35)$$

appears as a key tool to the solution of the constrained control problem and is central in the proposed positive invariance approach. It is the pole assignment procedure that fits well with this kind of problems. It is called inverse procedure by opposition to the direct one since the matrix H is now given and the feedback X or say F is to be determined. Consider the following constrained system:

$$\delta x(\cdot) = Ax(\cdot) + Bsat(u(\cdot)) \quad (1.36)$$

where

$$\delta x = \begin{cases} \dot{x}(t) & \text{for } CTC \\ x(k+1) & \text{for } DTC \end{cases}$$

and assume that:

- (AS1): The pair (A, B) is controllable.
- (AS2): The matrix A has $n - m$ stable eigenvalues.

We present, hereafter its detailed solution. In fact, the development below gives the necessary and sufficient condition of the existence of a non-trivial solution of (1.35) with the assumption (AS2), that is $\Lambda_2 \subset \sigma(A)$ where:

$$\Lambda_2 = \left\{\lambda_j \in \mathbb{C} / \lambda_j \neq 0,\ \lambda_j = \overline{\lambda_l},\ j, l = m+1, ..., n\right\}$$
the associated eigenvectors ξ_j are such that
$$A\xi_j = \lambda_j \xi_j, \quad j = m+1, ..., n.$$
$$\xi_{m+1}, ..., \xi_n \quad are \quad linearly \quad independent.$$

Give a diagonalizable matrix $H \in \mathbb{R}^{m\times m}$ satisfying the following assumptions:

$$\Lambda_1 = \sigma(H) = \left\{\lambda_i \in \mathbb{C}/\lambda_i \neq 0,\ \lambda_i = \overline{\lambda_l}\, i, l = 1, ..., m\right\} \tag{1.37}$$

$$\left.\begin{matrix} H\theta_i = \lambda_i \theta_i \\ B\theta_i \neq 0 \end{matrix}\right\} i = 1, ..., m \tag{1.38}$$

$$\theta_1, ..., \theta_m \quad are \quad linearly \quad independent \tag{1.39}$$

$$\Lambda_1 \cap \sigma(A) = \emptyset \tag{1.40}$$

The eigenvectors of the unknown matrix $A + BX$ are given by:

$$\begin{cases} \zeta_i = (\lambda_i I_n - A)^{-1} B\theta_i, & i = 1, ..., m \\ \zeta_j = \xi_j, \quad j = m+1, ..., n. \end{cases} \tag{1.41}$$

Theorem 1.6 ([6, 7]) *For a given matrix H satisfying (1.37)–(1.40), the following statements are equivalent:*

- *There exists a unique full rank solution F to equation (1.35) given by:*

$$F = \left[\theta_1\ \theta_2\ \dots\ \theta_m\ 0_{m+1}\ \dots\ 0_n\right]\left[\zeta_1\ \zeta_2\ \dots\ \zeta_n\right]^{-1} \tag{1.42}$$

- *Vectors ζ_i, $i = 1, ..., n$ given by (1.41) are linearly independent.*

The solution of Eq. (1.35) is needed to design admissible controllers for constrained both continuous-time and discrete-time systems (1.36). This solution is improved in [1, 21].

Remark 1.1 Without loss of generality, it was assumed that the system possesses $(n - m)$ stable eigenvalues. In fact, if this is not true, it is always possible to augment the representation as follows:

Let $v \in \mathbb{R}$ be a vector of fictitious inputs. Hence, systems (1.36) can be transformed as follows:

$$\delta x(\cdot) = Ax(\cdot) + \begin{bmatrix} B & 0 \end{bmatrix} \begin{bmatrix} u(\cdot) \\ v(\cdot) \end{bmatrix} \tag{1.43}$$

It should be noted that this augmentation limits the domain of linear behavior of the closed-loop system, but it is always possible to soften the fictitious limitations to enlarge the domain. Henceforth, for the obtained square system the problem of $(n - m)$ stable eigenvalues is eliminated and controllability is not affected.

1.4 Constrained State and Control

In this section, in addition to constraints on the control variables, constraints on the state are considered. In fact, in general cases, the system dynamics are nonlinear and to obtain a linear model state variables are limited within given sets. Here we present results related to such problems that is taking into account state together with control constraints in the phase of controllers synthesis. For presentation brevity, discrete-time and continuous-time systems are assembled in the same notations using the operator δ. Consider the linear system represented by:

$$\delta x(t) = Ax(t) + Bu(t) \tag{1.44}$$

where $x(t) \in \mathbb{R}^n$ and δ stands for $x(t+1)$ for discrete-time systems and $\dot{x}(t)$ for continuous-time systems. The state vector belongs to the polyhedral set of $\mathbb{R}^n$, containing the origin in its interior, defined by:

$$\mathscr{D}(G, \omega) = \{x \in \mathbb{R}^n / \ Gx \leq \omega, \ G \in \mathbb{R}^{g \times n} \omega \in \mathbb{R}_+^{g*} \ \ g \geq n, \ rank(G) = n\} \tag{1.45}$$

and $u(t) \in \Omega_1 \subset \mathbb{R}^m$ is the input vector constrained to evolve within the polyhedral set Ω_1 of $\mathbb{R}^m$ where:

$$\Omega_1 = \{u \in \mathbb{R}^m \ \ / \ \ u \ \ \leq \ \ \rho, \ \ \ \rho \in \mathbb{R}_+^{m*}\} \tag{1.46}$$

Using a linear state feedback control law

$$u(t) = Fx(t), \quad F \in \mathbb{R}^{m \times n} \tag{1.47}$$

such that all initial states within domain $\mathscr{D}(G, \omega)$ are transferred asymptotically to the origin while the control respects constraints (1.46).

The feedback control law (1.47) induces a domain of admissible states given by:

$$\mathscr{D}(F, \rho) = \{ \ x \in \mathbb{R}^n, \quad Fx \leq \rho\} \tag{1.48}$$

The property of control admissibility for all $x \in \mathscr{D}(G, \omega)$ can be expressed as the following polyhedral inclusion

$$\mathscr{D}(G, \omega) \subseteq \mathscr{D}(F, \rho) \tag{1.49}$$

If control constraints are respected, closed-loop behavior is linear and is given by:

$$\begin{aligned} \delta x(t) &= (A + BF)x(t) \\ &= A_{cl}x(t) \end{aligned} \tag{1.50}$$

The linear constrained regulator problem as stated above has a solution if and only if the assertions below hold true [26, 28]:

1 Matrix (A + BF) is asymptotically stable.
2 $\mathscr{D}(G, \omega)$ is positively invariant w.r.t. system (1.50).
3 $\mathscr{D}(G, \omega) \subseteq \mathscr{D}(F, \rho)$.

In order to formulate this problem as a linear programming problem, the Haar's lemma [16] is used:

Lemma 1.1 *Let $P \in \mathbb{R}^{n\times m}$, $Q \in \mathbb{R}^{l\times m}$ be real matrices, $p \in \mathbb{R}^n$ and $q \in \mathbb{R}^g$ be column vectors. For $\mathscr{D}(P, p) \neq \emptyset$, the following assertions are equivalent:*

$$\mathscr{D}(P, p) = \{x \in \mathbb{R}^m : Px \leq p\} \subset \mathscr{D}(Q, q) = \{x \in \mathbb{R}^m : Qx \leq q\}$$
$$\exists\ H \in \mathbb{R}^{l\times n} \textit{ with nonnegative entries such that } Q = HP \textit{ and } Hp \leq q.$$

This lemma translates the inclusion between polyhedral sets as well as the positive invariance conditions, to linear inequalities. These inequalities are useful to construct a linear programming algorithm that enables to compute the stabilizing state gain feedback F.

By virtue of the Lemma 1.1, it is easy to proof in a different way [17, 27] (for the discrete-time case) the necessary and sufficient condition of positive invariance given as follows [25]:

Proposition 1.1 *$\mathscr{D}(G, \omega)$ is positively invariant set w.r.t. motion of system (1.50) if and only if there exists a matrix $H \in \mathscr{M}_H$ such that:*

$$HG = G(A + BF) \tag{1.51}$$
$$\mathscr{H}\omega \leq 0. \tag{1.52}$$

where $\mathscr{M}_H$, $\mathscr{H}$ denote, respectively, the set of nonnegative matrices and $H - \mathbb{I}$ in the continuous-time case; whereas $\mathscr{M}_H$ in the DTC, $\mathscr{H}$ stand, respectively, for the set of matrices with nonnegative off diagonal elements and matrix H.

Remark 1.2 For the non-symmetrical case where the state domain of constraints is given by the following form:

$$\mathscr{D}(\mathscr{G}, \omega_1, \omega_2) = \{x \in \mathscr{R}^n, / \ -\omega_2 \leq \mathscr{G}x \leq \omega_1\ ;\ \omega_1, \omega_2 \in \mathbb{R}^l_+\} \tag{1.53}$$

can be written as the form (1.45), with:

$$G = \begin{bmatrix} \mathscr{G} \\ -\mathscr{G} \end{bmatrix} \text{ and } \omega = \begin{bmatrix} \omega_1 \\ \omega_2 \end{bmatrix} \tag{1.54}$$

1.5 Positive Invariance for Non Autonomous Systems

In this section, we will derive a result for the continuous-time case similar to that obtained by [3] for the discrete-time case. For this, let us consider the discrete-time system:

$$x(k+1) = Ax(k) + Bu(k), \tag{1.55}$$

where x represents the state vector in $\mathbb{R}^n$, u is the control constrained to lie in subset Ω given by (1.3), with $(m \leq n)$. A and B are matrices of appropriate sizes. Let us define also the state domain given by:

$$\mathscr{D}(\mathbb{I}, x_{min}, x_{max}) = \{x \in \mathbb{R}^n, -x_{min} \leq x \leq x_{max}, \quad x_{min}, x_{max} \in \mathbb{R}^n_+\} \tag{1.56}$$

Furthermore, we assume that:

$$(A, B) \text{ is stabilizable} \tag{1.57}$$

Then consider the set given by (1.56); the motion of the system (1.55) does not leave the domain $\mathscr{D}(\mathbb{I}, x_{min}, x_{max})$ for every initial condition $x_o \in \mathscr{D}(\mathbb{I}, x_{min}, x_{max})$ if and only if the conditions given in the following theorem hold true.

Theorem 1.7 ([2]) *Domain (1.56) (respectively $int\mathscr{D}(\mathbb{I}, x_{min}, x_{max})$) is positively invariant w.r.t. the motion of system (1.55) if and only if*

$$\tilde{A}_d\chi + \tilde{B}_d U \leqslant \chi, \textit{(respectively} < \chi) \tag{1.58}$$

where $\chi = \begin{bmatrix} x_{max} \\ x_{min} \end{bmatrix}$ *and* $U = \begin{bmatrix} u_{max} \\ u_{min} \end{bmatrix}$, $\tilde{A}_d$ $\tilde{B}_d$ *are defined by* (1.21)–(1.23).

Now, we introduce a lemma that connects the continuous-time case to the discrete one, such that one can easily extend condition (1.58) to the continuous-time case.

Let us consider the continuous-time system given by:

$$\dot{x}(t) = Ax(t) + Bu(t) \tag{1.59}$$

Lemma 1.2 *The domain $int\mathscr{D}(\mathbb{I}, x_{min}, x_{max})$ is positively invariant w.r.t. the motion of continuous-time system (1.59) if and only if it is positively invariant w.r.t. the motion of the approximating discrete-time system given by:*

$$x(k+1) = (\mathbb{I} + \tau A)x(k) + \tau Bu(k), \tag{1.60}$$

for some τ that satisfies:

$$0 < \tau < inf(1/|a_{ii}|), \forall i \in [1 \;\; n], \, such \, as \, a_{ii} < 0 \tag{1.61}$$

Remark 1.3 Lemma 1.2 is a restatement of Theorem 2.2 of [10]. Here, we give a different proof for completeness.

Proof:
Sufficiency: We assume that the domain $int\mathscr{D}(\mathbb{I}, x_{min}, x_{max})$ is positively invariant w.r.t to the motion of system (1.60) for some τ satisfying condition (1.61). Consider a current state $x(t) \in int\mathscr{D}(\mathbb{I}, x_{min}, x_{max})$, a Taylor development of the solution of the continuous-time system (1.59) is given by:

$$x(t+\tau) = x(t) + \tau(Ax(t) + Bu(t)) + \nu(\tau) \tag{1.62}$$

where $\nu/\tau \to 0$ when $\tau \to 0$. Let us denote $\xi(k) = x(t)$ and $v(k) = u(t)$ and the discrete-time system:

$$\xi(k+1) = (1 + \tau A)\xi(k) + \tau B v(k),$$

which admits $int\mathscr{D}(\mathbb{I}, x_{min}, x_{max})$ as a positively invariant set, hence the state $\xi(k+1) \in int\mathscr{D}(\mathbb{I}, x_{min}, x_{max})$. This allows us to write:

$$-x_{min} < \xi(k+1) < x_{max},$$

which is equivalent to:

$$-x_{min} < x(t) + \tau(Ax(t) + Bu(t)) < x_{max},$$

or similarly

$$-x_{min} + \nu(\tau) < x(t+\tau) < x_{max} + \nu(\tau),$$

Dividing by τ, one can write:

$$\frac{-x_{min} + \nu(\tau)}{\tau} < \frac{x(t+\tau)}{\tau} < \frac{x_{max} + \nu(\tau)}{\tau} \tag{1.63}$$

and considering the right-hand side of (1.63), there always exists $\alpha < 0$ such that:

$$\frac{x(t+\tau) - x_{max}}{\tau} - \frac{\nu(\tau)}{\tau} \leqslant \alpha < 0,$$

Or equivalently,

$$\frac{x(t+\tau) - x_{max}}{\tau} \leq \alpha + \frac{\nu(\tau)}{\tau},$$

and because $\frac{\nu(\tau)}{\tau}$ vanishes to 0 when $\nu \to 0$, we can always find $\nu_0 \in]0, \tau[$ such that:

$$\frac{x(t+\bar{\tau}) - x_{max}}{\bar{\tau}} \leq \alpha < 0, \ \forall \bar{\tau} \in]0, \tau_0[,$$

that is, $x(t+\bar{\tau}) < x_{max}$. The same reasoning for the left-hand side of (1.63) allows us to write:

$$x(t+\bar{\tau}) \in int\mathscr{D}(\mathbb{I}, x_{min}, x_{max}) \text{ for } \bar{\tau} > 0.$$

As a consequence, $(int\mathscr{D}(\mathbb{I}, x_{min}, x_{max}))$ is positively invariant with respect to the motion of the continuous-time system (1.59).

Necessity: Let us assume that the domain $int\mathscr{D}(\mathbb{I}, x_{min}, x_{max})$ is positively invariant w.r.t the motion of continuous-time system (1.59). Consider the non-quadratic and non-symmetric Lyapunov function [5] given by:

$$V(x) = \max\{\frac{x_i^+}{(x_{max})_i}, \frac{x_i^-}{(x_{min})_i}\} \tag{1.64}$$

For all $x \in int\mathscr{D}(\mathbb{I}, x_{min}, x_{max})$ we know that $\dot{V}(x) \leqslant 0$ which implies that [5]:

$$V(x + \varepsilon\dot{x}) \leqslant V(x), \ \forall \varepsilon > 0 \tag{1.65}$$

In particular, for $\varepsilon = \tau$

$$V(x + \tau\dot{x}) \leqslant V(x),$$

$$V(x + \tau(Ax + Bu)) \leqslant V(x),$$

$$V((\mathbb{I} + \tau A)x + \tau Bu) \leqslant V(x)$$

Which is equivalent to writing that:

$$V(x(k+1)) \leqslant V(x(k)),$$

respectively to the discrete-time system (1.60). This permits us to state that $V(x)$ is a Lyapunov function for system (1.60), and that the domain $int\mathscr{D}(\mathbb{I}, x_{min}, x_{max})$ is positively invariant w.r.t system (1.60), because it can be obtained as the set of $x \in \mathbb{R}^n$ such that $V(x) < 1$ [4]. □

In order to introduce an important result from the literature, we consider the continuous-time system (1.59) with a saturated state feedback:

$$u(t) = sat(Kx(t)), \ K \in \mathbb{R}^{m \times n}, \ rank\{K\} = m. \tag{1.66}$$

The saturation operator divides the state space into two regions:

(a) the polyhedral domain $\mathscr{D}(K, u_{max}, u_{min})$ given by:

$$\mathscr{D}(K, u_{max}, u_{min}) = \{x \in \mathbb{R}^n / -u_{min} \leqslant Kx \leqslant u_{max};\, u_{max}, u_{min} \in \mathbb{R}^m_+ - \{0\}\}, \tag{1.67}$$

where the system behavior in a closed loop is linear:

$$\dot{x}(t) = (A + BK)x(t). \tag{1.68}$$

(b) the domain $\mathbb{R}^n / \mathscr{D}(K, u_{max}, u_{min})$ where the closed-loop behavior is nonlinear. If one constrains the state to evolve into the domain $\mathscr{D}(K, u_{max}, u_{min})$ (i.e., the domain $\mathscr{D}(K, u_{max}, u_{min})$ is positively invariant), control saturation does not occur and linear behavior in the closed loop is guaranteed. Consequently, if the matrix K is chosen such that:

$$Re(\lambda_i) \leqslant 0, \ i = 1, ..., n \ , \tag{1.69}$$

where λ_i are the closed-loop eigenvalues, the asymptotic stability of system (1.59) with state feedback (1.66) is achieved. A necessary and sufficient condition of positive invariance of domain $\mathscr{D}(K, u_{max}, u_{min})$ is given below.

Now, we introduce a necessary and sufficient condition of positive invariance of domains of type $int\mathscr{D}(\mathbb{I}, x_{min}, x_{max})$ w.r.t system (1.59) in the open-loop.

Theorem 1.8 *Domain* $int\mathscr{D}(\mathbb{I}, x_{max}, x_{min})$ *is positively invariant w.r.t the motion of system* (1.59) *if and only if:*

$$\tilde{A}_c \chi + \tilde{B}_d U < 0. \tag{1.70}$$

where $\tilde{A}_c$ *and* $\tilde{B}_d$ *are defined by* (1.32) *and* (1.21), *respectively.*

Sufficiency: The domain $int\mathscr{D}(\mathbb{I}, x_{max}, x_{min})$ is positively invariant with the motion of the system (1.59) and (1.57) iff it is positively invariant w.r.t the motion of the approximating discrete-time system (1.60). According to the Theorem 1.7, this is equivalent to:

$$\tilde{\Gamma}_d(\tau)\chi + \tau \tilde{B}_d U < \chi, \tag{1.71}$$

where

$$\Gamma(\tau) = (\mathbb{I} + \tau A) \tag{1.72}$$

As τ satisfies the condition (1.61), one can write both quantities $(\mathbb{I} + \tau A)^+$ and $(\mathbb{I} + \tau A)^-$ as follows:

$$(\mathbb{I} + \tau A)^+ = \mathbb{I} + \tau A_1, \tag{1.73}$$

$$(\mathbb{I} + \tau A)^- = \mathbb{I} + \tau A_2, \tag{1.74}$$

Equation (1.71) becomes

$$\tau(\tilde{A}_c \chi + \tilde{B}_d U) < 0, \tag{1.75}$$

and sine τ is positive, this leads to:

$$\tilde{A}_c\chi + \tilde{B}_d U < 0, \tag{1.76}$$

Necessity. Assume that the condition (1.70) holds, then let us choose a scalar ε as follows:

$$0 < \varepsilon < inf(1/|a_{ii}|), \text{ for } a_{ii} < 0. \tag{1.77}$$

Multiplying the condition (1.70) by $\varepsilon > 0$ does not change the inequality, and we obtain

$$\varepsilon(\tilde{A}_c\chi + \tilde{B}_d U) < 0, \tag{1.78}$$

Adding the vector χ to both sides of the inequality allows us to write

$$\chi + \varepsilon(\tilde{A}_c\chi + \tilde{B}_d U) < \chi. \tag{1.79}$$

From above, it becomes obvious that:

$$\tilde{\Gamma}_d(\varepsilon) \leqslant \chi + \varepsilon\tilde{A}_c\chi < \chi - \varepsilon\tilde{B}_d U. \tag{1.80}$$

or equivalently

$$\tilde{\Gamma}_d(\varepsilon)\chi + \varepsilon\tilde{B}_d U < \chi, \tag{1.81}$$

and according to Theorem 1.7, this implies that the domain $int\mathscr{D}(\mathbb{I}, x_{max}, x_{min})$ is positively invariant w.r.t the discrete-time system given by:

$$x(k+1) = (\mathbb{I} + \varepsilon A)x(k) + \varepsilon Bu(k), \tag{1.82}$$

and by using Lemma 1.2, one may conclude that $int\mathscr{D}(\mathbb{I}, x_1, x_2)$ is positively invariant w.r.t the motion of the continuous-time system (1.59).

1.6 D-Positive Invariance

This section presents the extension of results of positive invariance without disturbances [24] to the case where the system is subject to additive bounded disturbances as it is a frequent situation in practice. These results can be seen as a direct application of results of the previous section.

Consider the linear time invariant autonomous system with additive bounded disturbances given by:

$$\delta z(t) = H\ z(t) + E\ p(t), \quad z(t_o) = z_o \tag{1.83}$$

where $z \in \mathbb{R}^m$ is the state, constrained to evolve in the domain:

$$\mathscr{D}_z = \{z \in \mathbb{R}^m / -z_{\min} \leq z(t) \leq z_{\max}, \quad z_{\min}, z_{\max} \in int\mathbb{R}^m_+\} \tag{1.84}$$

and $p(t)$ is the disturbance, bounded in the domain:

$$\mathscr{D}_P = \{p(t) \in \mathbb{R}^p / -p_{\min} \leq p(t) \leq p_{\max}, \quad p_{\min}, p_{\max} \in \mathbb{R}^p_+\} \tag{1.85}$$

Further, we denote:

$$\vartheta = \begin{bmatrix} z_{\max} \\ z_{\min} \end{bmatrix}; \varpi = \begin{bmatrix} p_{\max} \\ p_{\min} \end{bmatrix}$$

Now, we recall the definition of $\mathscr{D}_P$-positive invariance of domain $\mathscr{D}_z$, which is very useful for the sequel.

Definition 1.4 Domain $\mathscr{D}_z$ given by (1.84) is $\mathscr{D}_P$-positively invariant w.r.t. motion of system (1.83) if for all initial conditions $z_o \in \mathscr{D}_z$, the trajectory of the system $z(t, t_o, z_o) \in \mathscr{D}_z$ for all $p(t) \in \mathscr{D}_P, \quad t > t_o$.

$\mathscr{D}_P$-positive invariance conditions have already been reported in [3, 9] for the discrete-time case and in [23] for the continuous-time case. Let us recall these conditions hereafter:

Theorem 1.9 ([9, 23]) *Domain $\mathscr{D}_z$ is $\mathscr{D}_P$-positively invariant w.r.t. motion of the system with additive disturbances (1.83) if and only if matrix H satisfies:*

$$\tilde{H}_c \ \vartheta \ + \ \tilde{E}_d \ \varpi \ \leq \ 0 \text{ for the continuous-time case} \tag{1.86}$$
$$\tilde{H}_d \ \vartheta \ + \ \tilde{E}_d \ \varpi \ \leq \ \vartheta \text{ for the discrete-time case} \tag{1.87}$$

where $\tilde{H}_c$ and $\tilde{(\cdot)}_d$ are defined by (1.32) *and* (1.21)*, respectively.*

Evolution of the autonomous system (1.83) will respect constraints on the state $z(t)$ if domain $\mathscr{D}_z$ given by (1.84) is $\mathscr{D}_P$-positively invariant.

1.7 Saturated Control

In all the results presented above, saturation is avoided. In opposition, in what follows saturating control is permitted. Hence, modeling the obtained nonlinear system as a convex combination of a set of linear systems is the key to solve such cases [12, 19]. Further, stability conditions are revisited with this writing of the system. Consider the system:

$$\delta x(t) = Ax(t) + Bsat(u(t)) \tag{1.88}$$

with the state feedback:

$$u(t) = Fx(t), \quad F \in \mathbb{R}^{m \times n} \tag{1.89}$$

Define the subsets of $\mathbb{R}^n$,

$$\varepsilon(P, \rho) = \{x \in \mathbb{R}^n \mid x^T Px \leq \rho, \rho > 0\} \tag{1.90}$$

$$\mathscr{L}(F) = \{x \in \mathbb{R}^n \mid |F_j x| \leq 1,\ 1 \leq j \leq m\} \tag{1.91}$$

with P a positive definite matrix and F_j the jth row of the matrix $F \in \mathbb{R}^{m \times n}$. $\varepsilon(P, \rho)$ is an ellipsoid set while $\mathscr{L}(F)$ is a polyhedral set for which the saturation does not occur. Define matrices D_s as m by m diagonal matrices with elements either 1 or 0 and $D_s^- = \mathbb{I}_m - D_s$. There are 2^m possible combination with 1 and 0 leading to have 2^m different matrices D_s. Note that matrices D_s and D_s^- are introduced by [19] to model the nonlinear saturation function as a linear convex combination by using the following:

Lemma 1.3 ([19]) *For all $u \in \mathbb{R}^m$ and $v \in \mathbb{R}^m$ such that $|v_j| < 1$, $j \in [1, m]$*

$$sat(u) \in co\{D_s u + D_s^- v, s \in \mathscr{I}\} \tag{1.92}$$

with $\mathscr{I} = [1, \eta]$; $\eta = 2^m$ and $co\{.\}$ denotes the convex hull of $\{.\}$.

In this case, there exist $\gamma_1 \geq 0, \ldots, \gamma_\eta \geq 0$ satisfying $\sum_{s=1}^{\eta} \gamma_s = 1$ such that,

$$sat(u) = \sum_{s=1}^{\eta} \gamma_s [D_s u + D_s^- v] \tag{1.93}$$

If one uses a matrix G such that $v = Gx$, the system in closed loop becomes:

$$\delta x(t) = \sum_{s=1}^{\eta} \gamma_s(t) \left(A + B[D_s F + D_s^- G]\right) x(t) \tag{1.94}$$

The stability condition of the saturated system can be enunciated:

Theorem 1.10 ([18, 19]) *Given an ellipsoid $\varepsilon(P, \rho)$, if there exists a matrix $G \in \mathbb{R}^{m \times n}$ such that:*

- *for the discrete-time case:*

$$[A + B(D_i F + D_i^- G)]^T P [A + B(D_i F + D_i^- G)] - P \prec 0,\ \forall i \in [1, \eta], \tag{1.95}$$

- *for the continuous-time case:*

$$[A + B(D_i F + D_i^- G)]^T P + P[A + B(D_i F + D_i^- G)] \prec 0,\ \forall i \in [1, \eta], \tag{1.96}$$

and $\varepsilon(P, \rho) \subset \mathscr{L}(G)$, then $\varepsilon(P, \rho)$ is a contractively invariant set for the closed-loop system with saturation (1.94).

1.8 Singular Systems

In this section, we recall some definitions and results concerning the singular systems case. This class of systems with constrained control will be studied in a subsequent chapter of this book. Let us consider the singular system described by:

$$\begin{aligned} E\dot{x}(t) &= Ax(t) + Bsat(u(t)) \\ x(0) &= x_o \end{aligned} \tag{1.97}$$

where $x \in \mathbb{R}^n$ is the state, $u \in \mathbb{R}^m$ is the control with, $Rank(E) = r \leq n$. Assume that,

(AS1) (E, A, B) is stabilizable and $m \leq r$.

The control is assumed to be constrained in the set Ω given by (1.3).

In order to present some useful lemmas, matrices E and B are decomposed as follows:

$$E = [R \;\; 0] \, [S_0 \;\; S_\infty]^T, \quad S = [S_0 \;\; S_\infty] \tag{1.98}$$

$$B = [\Phi_0 \;\; \Phi_1] \, [Z_B^T \;\; 0]^T, \quad \Phi = [\Phi_0 \;\; \Phi_1] \tag{1.99}$$

where matrices Φ and S are orthogonal.

Definition 1.5 [14]

- The pair (E, A) is said to be regular if $det(sE - A)$ is not identically zero.
- The pair (E, A) is said to be impulse free if the degree of $det(sE - A) = rank(E)$
- The pair (E, A) is said to be admissible if it is regular, impulse free, and stable *i.e.,* the real part of all the finite generalized eigenvalues is negative.

Lemma 1.4 ([20]): *If the pencil $[E, A]$ is regular, then it has r finite eigenvalues if and only if:*

$$rank[E + AS_\infty S_\infty^T] = n. \tag{1.100}$$

Note that the pencil $[E, A]$ has r finite eigenvalues if it is impulse free, that is $deg(det(sE - A)) = rank(E) = r$. Further, if condition (1.100) holds, the pencil $[E, A]$ is regular.

The control is assumed here to be constrained in Ω given by (1.3). Generally, when the state is available, the control is given by:

$$u(t) = Fx(t), \quad F \in \mathbb{R}^{m \times n}, \quad rank(F) = m. \tag{1.101}$$

The unsaturated system in closed loop is then obtained by:

$$\begin{aligned} E\dot{x}(t) &= (A + BF)x(t) \\ &= A_{cl}x(t). \end{aligned} \tag{1.102}$$

With this system in closed loop, the induced constraint set on the state $\mathscr{D}(F, u_{max}, u_{min})$ is given by (1.10).

The gain matrix F is obtained by solving the following equation which was studied in [8]:

$$XA + XBXE = HXE. \tag{1.103}$$

The following development gives the necessary and sufficient condition of the existence of the non-trivial solution of (1.103) with the assumption that the pencil $[A, E]$ satisfies:

AS2) The pencil $[E, A]$ is regular and impulse free, *i.e.*, $rank[E + AS_\infty S_\infty^T] = n$. Give a diagonalizable matrix $H \in \mathbb{R}^{m\times m}$ satisfying the following assumptions:

$$\{\theta_1, \ldots, \theta_m\} \text{ are linearly independent} \tag{1.104}$$

$$\sigma(H) \bigcap \sigma([A, E]) = \emptyset \tag{1.105}$$

$$B\theta_i \neq 0, i = 1, \ldots, m \tag{1.106}$$

$$[\Xi_r \;\; S_\infty] \text{ is non singular} \tag{1.107}$$

where $\sigma(H)$ denotes the spectrum of H and $\sigma([A, E])$ denotes the generalized spectrum of the pencil $[A, E]$. Matrix $\Xi_r = [\xi_1, \ldots, \xi_r]$ with $\xi_i = (\lambda E - A)^{-1}B\theta_i$ while $\Theta_r = [\theta_1, \ldots, \theta_r]$.

Theorem 1.11 ([8]) *For given matrices A, B, E, H according to assumptions (AS1)-(AS2) and (1.104)–(1.106), there exists a solution of Eq. (1.103) if and only if:*

$$rank \begin{bmatrix} [E\Xi_r \;\; AS_\infty] \\ [\Theta_r \;\; O] \end{bmatrix} = rank\,[\Theta_r \;\; O] \tag{1.108}$$

In this case, all the solutions are given by:

$$X = [\Theta_r \;\; O]\Sigma^+ + Y(\mathbb{I} - \Sigma\Sigma^+) \tag{1.109}$$

where $\Sigma = [E\Xi_r \;\; AS_\infty]$, Σ^+ denotes any generalized inverse matrix of Σ satisfying $\Sigma\Sigma^+\Sigma = \Sigma$ and Y any arbitrary matrix of appropriate dimensions.

Corollary 1.1 *[8] For given matrices A, B, E, H according to assumptions (AS1)-(AS2) and (1.104)–(1.107), matrix F is given by:*

$$F = [\Theta_r \;\; O][\Xi_r \;\; S_\infty]^{-1} \tag{1.110}$$

Remark 1.4 According to Eq. (1.110) where matrix Θ_r is composed of r linearly independent eigenvectors of H, we have $rank(F) = rank(\Theta_r)$. It follows that matrix F is of full rank row if and only if $m \leq r$.

The definition of positive invariance is still the same as for non-singular systems.

Now, necessary and sufficient condition of positive invariance of the set

$$\mathscr{D}(F, u_{max}, u_{min})$$

given by (1.10) w.r.t system (1.102) can be presented by using a constrained feedback controller.

We assume in this subsection that the state is available. The following result of positive invariance of the set $\mathscr{D}(F, u_{max}, u_{min})$ w.r.t the system in closed loop (1.94) is stated.

Theorem 1.12 *The set $\mathscr{D}(F, u_{max}, u_{min})$ given by (1.10) is positively invariant w.r.t the system (1.102) if and only if there exists a matrix $H \in \mathbb{R}^{m\times m}$ such that,*

$$\Gamma A + \Gamma B \Gamma E = H \Gamma E, \tag{1.111}$$

$$\tilde{H}_c U \leq 0. \tag{1.112}$$

where $F = \Gamma E$

Proof:
See [7].

1.9 Other Lemmas

A useful lemma based on the pseudo inverse of a matrix is also recalled.

Lemma 1.5 ([22])*: The matrix system $M\,S = N$ has a solution in the variable S if and only if*

$$(\mathbb{I} - M\,M^{+})N = 0.$$

Moreover, all the solutions are given by,

$$S = M^{+}N + (\mathbb{I} - M^{+}M)L,$$

where L is an arbitrary matrix and M^{+} denotes the pseudo inverse of matrix M.

1.10 Conclusion

In this chapter, some preliminary results have been reminded for their usefulness in the next chapters. In order to avoid the repetition of certain number of lemmas and intermediate results in several chapters of this book, these lasts are gathered in this first chapter. Hence, it is preferable to start reading this book by studying this chapter.

References

1. A. Baddou, H. Maarouf, A. Benzaouia, Partial eigenstructure assignment problem and its application to the constrained linear problem. Int. J. Sys. Sci. **44**, 908–915 (2013)
2. A. Benzaouia, Application du concept d'invariance positive à l'étude des problèmes de commande des systèmes dynamiques discrets avec contraintes sur les entrées. Thesis of UPS LAAS, 88322, 1988
3. A. Benzaouia, C. Burgat, Regulator problem for linear discrete-time systems with non-symmetrical constrained control. Int. J. Control. **48**(6), 2441–2451 (1988)
4. A. Benzaouia, C. Burgat, Existence of non symmetrical Lyapunov functions for linear systems. Int. J. Syst. Sc. **20**, 560–597 (1989)
5. A. Benzaouia, A. Hmamed, Regulator problem for linear continuous-time systems with non-symmetrical constrained control. IEEE Trans. Aut. Control. **38**(10), 1556–1560 (1993)
6. A. Benzaouia, Resolution of equation XA+XBX=HX and the pole assignment problem. IEEE Trans. Aut. Control. **40**(10), 2091–2095 (1994)
7. A. Benzaouia, *Saturated switching systems*. LNC 426 (Springer, 2012). ISBN-13: 978-1447128991
8. A. Benzaouia, M. Darouach, A. Hmamed, Solution of equation XA + XBXE = HXE and the pole assignment for singular systems. IMA J. Math. Control. Inf. **29**(3), 343–356 (2012)
9. F. Blanchini, Feedback control for linear systems with state and control bounds in the presence of disturbance. IEEE Trans. Aut. Control. **35**, 1131–1135 (1990)
10. F. Blanchini, Feedback control for linear time invariant systems with states and control bounds in the presence of disturbances. IEEE Trans. Aut. Control. **35**, 1234 (1992)
11. F. Blanchini, Set Invariance in control. Automatica **35**, 1747–1767 (1999)
12. Y. Y. Cao, Z. Lin, Stability analysis of discrete-time systems with actuator saturation by saturation dependent Lyapunov function in *Proceedings of the 41th IEEE CDC*, Las Vegas, USA, (2002)
13. E.B. Chegancas, S. Tarbouriech, On positive invariance and output feedback stabilization of input constrained linear systems in *Proceeding of the ACC*, Baltimore, U.S.A., (1994), pp. 2740–2744,
14. L. Dai, *Lecture Notes in Control and Information Sciences, in Singular Control Systems* (Springer, Berlin, 1989)
15. P.O. Gutman, P. Hagander, A new design of constrained controllers for linear systems. IEEE Trans. Aut. Control. **30**, 22–33 (1985)
16. A. Haar, *Über lineare Ungleichungen*. 1918, Reprinted in: Alfred Haar, Gesammelte Arbienten (Akadémai Kiadó, Bubapest 1959)
17. J.C. Hennet, *Une extension du lemme de Farkas et son application au problème de régulation linéaire sous constraintes*. Comptes-rendus de l'Académie des Sciences, 308, Série I, 415–419 (1989)
18. T. Hu, Z. Lin, *The equivalence of several set invariance conditions under saturation in Proceeding of the 41th IEEE CDC* (Las Vegas, USA, 2002)
19. T. Hu, Z. Lin, B.M. Chen, Analysis and design for discrete-time linear systems subject to actuator saturation. Syst. Control. Lett. **45**, 97–112 (2002)
20. J. Kaustky, N.K. Nichols, E.K.W. Chu, Robust pole assignment in singular systems. Linear Algebr. Appl. **121**, 9–37 (1989)
21. H. Maarouf, The resolution of the equation XA+XBX=HX and the pole assignment problem: a general approach. Automatica **79**, 162–66 (2017)
22. J. R. Magnus, H. Neudecker, *Matrix differential calculus with application in statistic and econometrics*, (Wiley, 1988)
23. F. Mesquine, D. Mehdi, Constrained observer based controller for linear continuous time systems. Int. J. Syst. Sci. **27**(12), 1363–1369 (1996)
24. F. Mesquine, F. Tadeo, A. Benzaouia, Regulator problem for linear systems with constrained control and its increment in *Proceedings of the 15th IFAC*, Barcelona, Spain, (2002)

25. F. Mesquine, A. Benlemkadem, Robust linear constrained regulator problem. IMA J. Math. Control. Inf. (2006). doi:10.1093/imamci/dn1015
26. B.E.A. Milani, A.N. Carvalho, Robust linear regulator design for discrete-time systems under polyhedral constraints. Automatica **31**, 1489–1493 (1995)
27. A.A. Ten Dam, J.W. Nieuwenhuis, A linear programming algorithm for invariant polyhedral sets of discrete-time linear systems. Syst. Control. Lett. **25**, 337–341 (1995)
28. M. Vassilaki, G. Bitsoris, Constrained regulation of linear continuous-time dynamical systems. Syst. Control. Lett. **13**, 247–252 (1989)
29. M. Vassilaki, J.C. Hennet, G. Bitsoris, Feedback control of linear discrete-time system under state and control constraints. Int. J. Control. **47**, 1727–1735 (1988)

Chapter 2
Robust Constrained Linear Regulator Problem

2.1 Introduction

Generally, for physical, technological, and/or security reasons, any physical system is subject to functional limitations, and moreover, the adopted models are often subject to uncertainties. These uncertainties may find their origin in the modeling and measurement errors or on the computation approximations (see [1–5] and the references therein). As a consequence, the simultaneous presence of uncertainties and constraints in general physical systems has interested many authors to combine the techniques of robust control and constrained control [6, 7]. With the same objective and as an extension to the uncertain case of [8], the first part of this chapter is devoted to design and study robust-constrained regulators. Necessary and sufficient conditions of positive invariance of polyhedral domains with respect to motion of uncertain continuous-time systems are derived.

Further, one has to concede that there are several approaches proposed in the literature to solve this problem [6, 8–15] and others. From these approaches, the positive invariance approach is selected in the second part to calculate robust constant state feedback controllers [6, 10, 16, 17]. But from a different point of view for computation techniques in the design of such controllers, the idea is to translate the results found within the framework of positive invariance to algorithms of linear programming [18–20]. To this end, necessary and sufficient conditions of positive invariance are re-formulated in this part. This enables one to use linear programming algorithms to find robust-constrained regulators for both continuous-time and discrete-time linear systems.

A. Benzaouia et al., *Saturated Control of Linear Systems*, Studies in Systems, Decision and Control 124, https://doi.org/10.1007/978-3-319-65990-9_2

2.2 Robust-Constrained Linear Control

2.2.1 Problem Statement

Consider the linear uncertain continuous-time system represented by the following state space model:

$$\dot{x}(t) = A(q_A(t))x(t) + B(q_B(t))u(t) \tag{2.1}$$

where $x(t) \in \mathbb{R}^n$ denotes the state vector, $u(t) \in \Omega \subset \mathbb{R}^m$ is the control input. The control input is restricted (by saturation) to evolve in the following polyhedral set:

$$\Omega = \{u(t) \in \mathbb{R}^m / -u_{\min} \leq u \leq u_{\max};\ u_{\min}, u_{\max} \in int\mathbb{R}^m_+\} \tag{2.2}$$

$q_A(t) \in \Gamma_A \subset \mathbb{R}^{p_A}$ (resp. $q_B(t) \in \Gamma_B \subset \mathbb{R}^{p_B}$) is the uncertain vector. Γ_A and Γ_B are compact convex sets including the origin in their interiors. These vectors $q_A(t)$ and $q_B(t)$ measure the uncertainty in the model, affecting the parameters of the matrices A and B as follows:

$$\begin{aligned} A(q(t)) &= A_o + \sum_{h=1}^{p_A} A_h q_{Ah}(t) \\ B(q(t)) &= B_o + \sum_{h=1}^{p_B} B_h q_{Bh}(t) \end{aligned} \tag{2.3}$$

where $q_{Ah}(t)$ and $q_{Bh}(t))$ represent the hth component of vectors $q_A(t)$ and $q_B(t)$, respectively:

$$\begin{aligned} q_A(t) &= [q_{A1}(t)\ \ q_{A2}(t)\ \ \ldots\ \ q_{Ap_A}(t)]^T \\ q_B(t) &= [q_{B1}(t)\ \ q_{B2}(t)\ \ \ldots\ \ q_{Bp_B}(t)]^T \end{aligned} \tag{2.4}$$

Convexity and compactness of the set Γ_A imply that there exist μ_A vertices $\nu_i, i = 1, \ldots, \mu_A$ of Γ_A, such that every $q_A \in \Gamma_A$ can be written as a convex combination of ν_i as:

$$q_A = \sum_{i=1}^{\mu_A} \alpha_i \nu_i \ \text{ with } \sum_{i=1}^{\mu_A} \alpha_i = 1,\ 0 \leq \alpha_i \leq 1\ ,\ \text{ for } i = 1, \ldots, \mu_A \tag{2.5}$$

The set Γ_B is also convex and compact, so there also exist μ_B vertices $\nu_j, j = 1, \ldots, \mu_B$ of Γ_B such that every $q_B \in \Gamma_B$ can be written as a convex combination of ν_j as:

$$q_B = \sum_{j=1}^{\mu_B} \beta_j \nu_j \text{ with } \sum_{j=1}^{\mu_B} \beta_j = 1, \; 0 \leq \beta_j \leq 1 \;, \text{ for } j = 1, \ldots, \mu_B, \tag{2.6}$$

and the matrix $A(q_A(t))$ (resp. $B(q_B(t)))$ as:

$$A(q_A(t)) = \sum_{i=1}^{\mu_A} \alpha_i(t) \; A_i \quad \left(\text{resp. } B(q_B(t)) = \sum_{j=1}^{\mu_B} \beta_j(t) \; B_j \right), \tag{2.7}$$

with $A_i = A(\nu_i)$ (resp. $B_j = B(\nu_j)$).

Assume that the pair $(A(q_A(t)), B(q_B(t))$ is controllable for every $q_A \in \Gamma_A$ and $q_B \in \Gamma_B$.

The nominal system is given by:

$$\dot{x}(t) = A_o x(t) + B_o u(t). \tag{2.8}$$

The robust-constrained regulator problem, which will be studied in this chapter, is to find a state feedback control law:

$$u(t) = \; Fx(t), \;\; F \in \mathbb{R}^{m \times n} \text{ with } rank(F) \; = m, \tag{2.9}$$

which stabilizes the nominal system

$$\dot{x}(t) = (A_o + B_o F) x(t), \tag{2.10}$$

respecting the control constraints (2.2), such that the feedback system is robustly stable against parametric uncertainty: That is, the feedback stabilizes also the uncertain system (2.1). Application of this control law, while respecting control constraints, leads to the closed-loop system:

$$\begin{aligned}
\dot{x}(t) &= (A(q_A(t)) + B(q_B(t))F)\, x(t) && (2.11)\\
&= \sum_{j=1}^{\mu_B} \sum_{i=1}^{\mu_A} \beta_j \alpha_i \left(A_i + B_j F\right) && (2.12)\\
&= \sum_{j=1}^{\mu_B} \sum_{i=1}^{\mu_A} \beta_j \alpha_i \; A_{cl}(\nu_{ij}) && (2.13)\\
&= A_{cl}(q(t)) && (2.14)
\end{aligned}$$

where $q(t) \in \Gamma = (\Gamma_A \times \Gamma_B)$, and ν_{ij} denotes the vertices of Γ $(i = 1, ..., \mu_A; j = 1, ..., \mu_B)$.

Also, the proposed control law induces the following set of linear behavior in state space:

$$\mathscr{D} = \{x \in \mathbb{R}^n / \; -u_{\min} \le Fx \le u_{\max}, \; u_{\min}, u_{\max} \in int\mathbb{R}^m_+\} \tag{2.15}$$

Note here that as long as the system states remain in the domain $\mathscr{D}$, the linear behavior is guaranteed. Otherwise, the closed-loop system is given by:

$$\dot{x}(t) = A(q_A(t))\,x(t) + B(q_B(t))sat(F\,x(t)) \tag{2.16}$$

It is worth noticing that positive invariance of domain $\mathscr{D}$ given by (2.15), with respect to (w.r.t.) motion of the closed-loop system (2.11), is the cornerstone to derive robust-constrained regulators.

2.2.2 Design of Robust Controller

This section is devoted to the extension of the preliminary results presented in Chap. 1 to the case of uncertain plants. This extension gives conditions for designing robust controllers in the presence of uncertainty and input limitations. The following lemma is fundamental to extend necessary and sufficient conditions for the positive invariance of the polyhedral domain:

$$\mathscr{D}_p = \{z \in \mathbb{R}^m / -p_2 \le z \le p_1, \; p_1, p_2 \in int\mathbb{R}^m_+\} \tag{2.17}$$

with respect to motion of the autonomous uncertain system:

$$\dot{z}(t) = H(q(t))z(t) \tag{2.18}$$

Lemma 2.1 *The set $\mathscr{D}_p$ given by (2.17) is positively invariant with respect to motion of system (2.18) if and only if it is positively invariant with respect to motion of system (2.18) at vertices v_{ij}, $i = 1, ..., \mu_A$, $j = 1, ..., \mu_B$ of the set Γ.*

Proof

Necessity: It is obvious that if the domain is positively invariant for every q, it is especially positively invariant at the vertices v_{ij}, $i = 1, ..., \mu_A$, $j = 1, ..., \mu_B$.

Sufficiency: Suppose that $\mathscr{D}_P$ is positively invariant with respect to motion of system (2.18) for $q = v_{ij}$, $i = 1, ..., \mu_A$, $j = 1, ..., \mu_B$ then:

$$\tilde{H}_c(v_{ij})\varpi \le 0 \text{ with } \varpi = [p_1^T \; p_2^T]^T,$$

decomposing this inequality, it is possible to write:

$$H_1(v_{ij})\ p_1 + H_2(v_{ij})\ p_2 \leq 0, \tag{2.19}$$

$$H_2(v_{ij})\ p_1 + H_1(v_{ij})\ p_2 \leq 0, \tag{2.20}$$

however, according to the definition of matrices H_1 and H_2 given by (1.33), inequality (2.19) can be written as, for $l = 1, ..., m$,

$$h_{ll}(v_{ij})p_1^l + \sum_{\substack{k=1\\k\neq l}}^{m} h_{lk}^{+}(v_{ij})p_1^k + \sum_{\substack{k=1\\k\neq l}}^{m} h_{lk}^{-}(v_{ij})p_2^k \leq 0,$$

multiplying every inequality by α_i, β_j $i = 1, ..., \mu_A, j = 1, ..., \mu_B$, and summing on i and j, leads to

$$\sum_{j=1}^{\mu_B}\sum_{i=1}^{\mu_A} \beta_j\alpha_i h_{ll}(v_{ij})p_1^l + \sum_{j=1}^{\mu_B}\sum_{i=1}^{\mu_A}\sum_{\substack{k=1\\k\neq l}}^{m} \beta_j\alpha_i h_{lk}^{+}(v_{ij})p_1^k + \sum_{j=1}^{\mu_B}\sum_{i=1}^{\mu_A}\sum_{\substack{k=1\\k\neq l}}^{m} \beta_j\alpha_i h_{lk}^{-}(v_{ij})p_2^k \leq 0,$$

which is equivalent to write

$$\sum_{j=1}^{\mu_B}\sum_{i=1}^{\mu_A} \beta_j\alpha_i\ h_{ll}(v_{ij})\ p_1^l + \sum_{\substack{k=1\\k\neq l}}^{m}\sum_{j=1}^{\mu_B}\sum_{i=1}^{\mu_A} \beta_j\alpha_i\ h_{lk}^{+}(v_{ij})\ p_1^k + \sum_{\substack{k=1\\k\neq l}}^{m}\sum_{j=1}^{\mu_B}\sum_{i=1}^{\mu_A} \beta_j\alpha_i\ h_{lk}^{-}(v_{ij})\ p_2^k \leq 0,$$

using the definitions of h_{ij}^{+} and h_{ij}^{-} given in (1.33), it is possible to see that:

$$\left(\sum_{j=1}^{\mu_B}\sum_{i=1}^{\mu_A} \beta_j\ \alpha_i\ h_{lk}(v_{ij})\right)^{+} \leq \sum_{j=1}^{\mu_B}\sum_{i=1}^{\mu_A} \beta_j\ \alpha_i\ h_{lk}^{+}(v_{ij}),$$

and

$$\left(\sum_{j=1}^{\mu_B}\sum_{i=1}^{\mu_A} \beta_j\ \alpha_i\ h_{lk}(v_{ij})\right)^{-} \leq \sum_{j=1}^{\mu_B}\sum_{i=1}^{\mu_A} \beta_j\ \alpha_i\ h_{lk}^{-}(v_{ij}),$$

then:

$$\sum_{j=1}^{\mu_B}\sum_{i=1}^{\mu_A} \beta_j\ \alpha_i h_{ll}(v_{ij})p_1^l + \sum_{\substack{k=1\\k\neq l}}^{m} \left(\sum_{j=1}^{\mu_B}\sum_{i=1}^{\mu_A} \beta_j\ \alpha_i h_{lk}(v_{ij})\right)^{+}$$

$$p_1^k + \sum_{\substack{k=1\\k\neq l}}^{m} \left(\sum_{j=1}^{\mu_B}\sum_{i=1}^{\mu_A} \beta_j\ \alpha_i\ h_{lk}(v_{ij})\right)^{-} p_2^k \leq 0.$$

According to (2.5) and (2.6) the following result is obtained:

$$H(q(t)) = \sum_{j=1}^{\mu_B} \sum_{i=1}^{\mu_A} \beta_j \, \alpha_i \, H(\nu_{ij}),$$

therefore,

$$h_{ll}(q(t)) \, p_1^l + \sum_{\substack{k=1 \\ k \neq l}}^{m} h_{lk}^{+}(q(t)) \, p_1^k + \sum_{\substack{k=1 \\ k \neq l}}^{m} h_{lk}^{-}(q(t)) \, p_2^k \leq 0,$$

hence,

$$H_1(q(t)) \, p_1 + H_2(q(t)) \, p_2 \leq 0.$$

Using a similar reasoning, it is possible to prove that (2.20) leads to:

$$H_2(q(t)) \, p_1 + H_1(q(t)) \, p_2 \leq 0,$$

which gives the sufficient condition:

$$\widetilde{H}_c(q(t)) \, \varpi \leq 0.$$

Consequently, the set $\mathscr{D}_P$ is positively invariant with respect to motion of system (2.18) $\forall \, q(t) \in \Gamma$. □

Now, it is possible to make the desired extension. Hence, let us consider the non-autonomous uncertain system (2.1) with constrained control (2.2). The application of the feedback control law (2.9) leads to the closed-loop system (2.11).

Theorem 2.1 *The subset $\mathscr{D}$ given by (2.15) is positively invariant with respect to motion of the uncertain system (2.11) with constrained control (2.2) if and only if there exist matrices $H(\nu_{ij})$ for $i = 1, ..., \mu_A$, $j = 1, ..., \mu_B$ such that:*

$$F A_{cl}(\nu_{ij}) = H(\nu_{ij}) \, F \tag{2.21}$$

$$\widetilde{H}_c(\nu_{ij}) \, U \leq 0 \tag{2.22}$$

Proof

Sufficiency: Assume that conditions (2.21) and (2.22) hold. By virtue of Theorem 1.4, the set $\mathscr{D}$ is positively invariant with respect to motion of system (2.11) at the vertices ν_{ij} for $i = 1, ..., \mu_A$, $j = 1, ..., \mu_B$. Further, consider the following change of coordinates

$$z(t) = F x(t) \tag{2.23}$$

the system (2.11) becomes system (2.18), and $\mathscr{D}$ is transformed as follows:

$$\mathscr{D}_z = \{z \in \mathbb{R}^m / -u_{\min} \le z \le u_{\max}, u_{\min}, u_{\max} \in \mathbb{R}_+^m\}.$$

Hence, $\mathscr{D}_z$ is positively invariant with respect to motion of system (2.18) at the vertices v_{ij}, for $i = 1, ...\mu_A, i = 1, ..., \mu_B$. Bearing in mind Lemma 2.1, $\mathscr{D}_z$ is positively invariant with respect to system (2.18) $\forall q \in \Gamma$. Consequently, $\mathscr{D}$ is positively invariant with respect to system (2.11) $\forall q \in \Gamma$.

Necessity: Let $\mathscr{D}$ be positively invariant with respect to system (2.11), so it is positively invariant at the vertices by using Lemma 2.1. Therefore, by virtue of Theorem 1.4, there exist matrices $H(v_{ij})$ satisfying conditions (2.21) and (2.22) for every $i = 1, ..., \mu_A, j = 1, ..., \mu_B$. □

Remark 2.1 For the sequel, and without loss of generality, assume that the system is square, i.e., $n = m$. In fact, if the system has n state and m input $m \le n$, then the system is augmented with constrained fictitious inputs v as presented above [6]. The resulting system is then given by:

$$\dot{x}(t) = A(q(t))x(t) + B_a(q(t)) \begin{bmatrix} u(t) \\ v(t) \end{bmatrix} \tag{2.24}$$

$B_a(q(t))$ is the matrix $B(q(t))$ augmented by $(n - m)$ null columns.

Corollary 2.1 *The positive invariance of domain $\mathscr{D}$ with respect to motions of system (2.24) with state feedback matrix F satisfying condition (2.9) implies the asymptotic stability of system (2.24) for every $x(0) \in \mathscr{D}$.*

Proof

Assume that domain $\mathscr{D}$ is positively invariant with respect to motion of system (2.24), then by virtue of the Theorem 2.1 and using the Lemma 2.1, we have:

$$FA_{cl}(q(t)) = H(q(t))\, F \tag{2.25}$$

$$\tilde{H}_c(q(t))\, U \le 0 \tag{2.26}$$

then using Theorem 2.1, the non-symmetric non-quadratic function

$$V(x) = \underset{l}{Max}\ max \left\{ \frac{(Fx)_l^+}{u_{\max}^l}, \frac{(Fx)_l^-}{u_{\min}^l} \right\}$$

is a common Lyapunov function of all stationary configuration of the uncertain system (2.11) for every $q \in \Gamma$. Hence, the system (2.11) is asymptotically stable. □

2.3 Robust Control with Constrained State and Input

In this second part, the robustness problem for constrained states and control is addressed. Different writing of the constraints is used. Further, conditions are expressed as linear programming problem.

2.3.1 Problem Statement

Let us consider the linear system represented by:

$$\delta x(t) = Ax(t) + Bu(t) \tag{2.27}$$

where $x(t) \in \mathbb{R}^n$ is the state vector belonging to the polyhedral set of $\mathbb{R}^n$, containing the origin in its interior, defined by :

$$\mathscr{D}(G, \omega) = \{x \in \mathbb{R}^n / \; Gx \leq \omega, \;\; G \in \mathbb{R}^{g \times n}, \;\; \omega \in \mathbb{R}_+^{g*}, \;\; g \geq n, \;\; rank(G) \; = \; n\} \tag{2.28}$$

and $u(t) \in \Omega_1 \subset \mathbb{R}^m$ is the input vector constrained to evolve within the polyhedral set Ω_1 of $\mathbb{R}^m$ where,

$$\Omega_1 = \{u \in \mathbb{R}^m \;\;\; / \;\;\; u \;\;\; \leq \;\;\; \rho, \;\;\;\; \rho \in \mathbb{R}_+^{m*}\} \tag{2.29}$$

The linear-constrained regulator problem (LCRP) can be stated as: Find a linear state feedback control law given by

$$u(t) = Fx(t), \quad\;\; F \in \mathbb{R}^{m \times n}, \tag{2.30}$$

such that all initial states within domain $\mathscr{D}(G, \omega)$ are transferred asymptotically to the origin while the control respects constraints (2.29).

The feedback control law (2.30) induces a domain of admissible states given by

$$\mathscr{D}(F, \rho) = \{ \;\; x \in \mathbb{R}^n, \quad\;\; Fx \leq \rho\} \tag{2.31}$$

The property of control admissibility for all $x \in \mathscr{D}(G, \omega)$ can be expressed as the following polyhedral inclusion

$$\mathscr{D}(G, \omega) \subseteq \mathscr{D}(F, \rho) \tag{2.32}$$

If control constraints are respected, closed-loop behavior is linear and is given by

$$\begin{aligned} \delta x(t) &= (A + BF)x(t) \\ &= A_{cl}x(t) \end{aligned} \tag{2.33}$$

Let us now consider that the system parameters are not perfectly known. The linear uncertain system has the following form:

$$\delta x(t) \;=\; A(q_A(t))\, x(t) \;+\; B(q_B(t))\, u(t), \tag{2.34}$$

where $q_A(t) \in \Gamma_A \subset \mathbb{R}^{p_A}$, and $q_B(t) \in \Gamma_B \subset \mathbb{R}^{p_B}$ where Γ_A and Γ_B are compact convex sets including the origin in their interiors. Uncertainties affect the system parameters as detailed in (2.3) and (2.4).

The Robust Linear-Constrained Regulator Problem (RLCRP), as studied hereafter, consists to find a robust regulator that transfers asymptotically all initial states in $\mathscr{D}(G, \omega)$ to the origin while the state and control respect constraints (2.28) and (2.29). The closed-loop system becomes :

$$\begin{aligned}\delta x(t) &= [A(q_A(t)) + B(q_B(t))F]\, x(t) \\ &= A_{cl}(q(t))x(t)\end{aligned} \tag{2.35}$$

where $q(t) \in \Gamma = \Gamma_A \times \Gamma_B$. The vertices of Γ are noted v_{ij} for $i = 1, ..., \mu_A$ and $j = 1, ..., \mu_B$.

Remark 2.2 Without loss generality, it was assumed that $rank(G) = n$. In fact, if this condition is not satisfied, not penalizing fictitious constraints on the state vector can be imposed. For example, consider that $n = 2$ and that the constraints on the state vector are given by:

$$\begin{bmatrix} g_{11} & g_{12} \\ g_{21} & g_{22} \end{bmatrix} \begin{bmatrix} x_1 \\ x_2 \end{bmatrix} \leq \begin{bmatrix} \omega_1 \\ \omega_2 \end{bmatrix}$$

and we have $rank(G) = 1$. We impose fictitious suitable constraints, in order to have $rank(G) = 2$, as follows

$$\begin{bmatrix} g_{11} & g_{12} \\ g_{21} & g_{22} \\ g_{a1} & g_{a2} \end{bmatrix} \begin{bmatrix} x_1 \\ x_2 \end{bmatrix} \leq \begin{bmatrix} \omega_1 \\ \omega_2 \\ \omega_a \end{bmatrix}$$

where the row g_a of matrix G and the constraint ω_a are chosen to be not penalizing for the initial problem. It is worth to note here that in [19, 21] for the discrete-time case, the matrix G is required to be square and invertible, which is very restrictive in our sense.

Now, the extension of positive invariance conditions for a polyhedral domain with respect to uncertain systems is studied. Necessary and sufficient conditions are established:

Lemma 2.2 *The set $\mathscr{D}(G, \omega)$, given by (2.28), is positively invariant with respect to motion of system (2.35) if and only if it is positively invariant with respect to motion of system (2.35) at vertices v_{ij} for $i = 1, ..., \mu_A$, $j = 1, ..., \mu_B$.*

Proof
Similar to proof of Lemma 2.1. □

Now, with this background, the main result of this section can be worked out.

2.3.2 Robust-Constrained Regulator Problem

As a first step, in this section, conditions to find a solution to the LCRP are formulated as linear inequalities. In a second step, these conditions are extended to the case of uncertain systems. This extension allows to find a solution to the RLCRP. The problem is cast into a linear programming feasibility formulation. Further, the introduction of an objective optimization variable ε enables to solve a linear programming-based algorithm giving stabilizing robust regulators respecting the state and control constraints (2.28) and (2.29).

Proposition 2.1 *A matrix F is solution to the LCRP if there exist matrices $H \in \mathscr{M}_H \subset \mathbb{R}^{g\times g}$, and $T \in \mathbb{R}^{m\times g}$ with nonnegative components ($T \geq 0$) such that:*

$$HG = G(A + BF) \tag{2.36}$$

$$\mathscr{H}\omega < 0 \tag{2.37}$$

$$F = TG \tag{2.38}$$

$$T\omega \leq \rho \tag{2.39}$$

where for a matrix $H \in \mathscr{M}_H$, we use $\mathscr{H}$ to note : $H - \mathbb{I}$ in the DTC, and H in the CTC. Further, the set $\mathscr{M}_H$ denotes the set of nonnegative matrices for the DTC and the set of matrices with nonnegative off-diagonal elements for the CTC (i.e., matrix H is Metzler, such that $h_{ij} \geq 0\ \forall i \neq j$).

Proof
Assume that conditions (2.36) and (2.37) hold true. By virtue of Proposition 1.1, the domain $\mathscr{D}(G, \omega)$ is positively invariant with respect to motion of system (2.33). The existence of a matrix T, such that conditions (2.38) and (2.39) hold and using the Haar's lemma, leads to the implication $Gx(t) \leq \omega \Rightarrow Fx(t) \leq \rho$ which is equivalent to:

$$\mathscr{D}(G, \omega) \subseteq \mathscr{D}(F, \rho) \tag{2.40}$$

Hence, the closed-loop behavior is linear and from the existence of a matrix H that satisfies (2.36) and (2.37), one can prove that the non-quadratic function:

$$V(x(t)) = \underset{i}{Max}\left\{\frac{|(Gx(t))_i|}{\omega_i}\right\}, \tag{2.41}$$

is a Lyapunov function for the system. Then, the asymptotic stability is guaranteed (for details about the Lyapunov function, see demonstration of Proposition 2.2). Finally, F is solution to the LCRP. □

Consider now that the system parameters are uncertain. As previously stated, the closed-loop uncertain system is given by (2.35).

Proposition 2.2 *A matrix* $F \in \mathbb{R}^{m\times n}$ *is solution to the RLCRP if there exist matrices* $H_{ij} \in \mathscr{M}_H \subset \mathbb{R}^{g\times g}$, $i = 1, ..., \mu_A$, $j = 1, ..., \mu_B$, *and* $T \in \mathbb{R}^{m\times g}$ *with nonnegative components* ($T \ \geq \ 0$) *satisfying conditions*

$$H_{ij}G = G(A_i + B_jF) \tag{2.42}$$

$$\mathscr{H}_{ij}\omega < 0 \tag{2.43}$$

$$F = TG \tag{2.44}$$

$$T\omega \leq \rho \tag{2.45}$$

Proof

First, using the Haar's lemma, conditions (2.44) and (2.45) imply the inclusion (2.40) guaranteeing the admissibility, and hence, the state and control constraints are not violated, and the closed-loop model is always linear.

Further, consider that the conditions (2.42) and (2.43) hold. By virtue of Proposition 1.1, the domain $\mathscr{D}(G, \omega)$ is positively invariant with respect to motion of system (2.35) at the vertices v_{ij} for $i = 1, ..., \mu_A$, $j = 1, ... \mu_B$. Using the Lemma 2.2, the domain is also positively invariant for every $q \in \Gamma$, which enables to write that there exist matrices $H(q(t)) \in \mathscr{M}_H$ such that:

$$H(q(t))G = GA_{cl}(q(t)), \tag{2.46}$$

bearing in mind that:

$$H(q(t)) := \sum_{i=1}^{\mu_A}\sum_{j=1}^{\mu_B} \alpha_i(t)\beta_j(t)H_{ij}, \tag{2.47}$$

one can conclude that:

$$\mathscr{H}(q(t))\omega \leq 0. \tag{2.48}$$

For the discrete-time case, consider the non-quadratic function given by:

$$V(x(t)) = \underset{i}{Max}\left\{\frac{|(Gx(t))_i|}{\omega_i}\right\}. \tag{2.49}$$

This function is positive definite. Let us compute its rate of increase along the trajectories of our system

$$\begin{aligned}\Delta V(x(t)) &= V(x(t+1)) - V(x(t)) \\ &= \underset{i}{Max}\left\{\frac{|(GA_{cl}(q(t))x(t))_i|}{\omega_i}\right\} - \underset{i}{Max}\left\{\frac{|(Gx(t))_i|}{\omega_i}\right\} \\ &= \underset{i}{Max}\left\{\frac{|(GA_{cl}(q(t))x(t))_i|}{\omega_i} - \underset{i}{Max}\left\{\frac{|(Gx)_i|}{\omega_i}\right\}\right\}.\end{aligned}$$

According to (2.46),

$$\Delta V(x(t)) = \underset{i}{Max}\left\{\frac{|(H(q(t))Gx(t))_i|}{\omega_i} - \underset{i}{Max}\left\{\frac{|(Gx(t))_i|}{\omega_i}\right\}\right\},$$

$|(H(q(t))Gx(t))_i|$ can be raised as follows:

$$|(H(q(t))Gx(t))_i| = \left|\sum_{j=1}^{g} h_{ij}(q(t))\ (Gx(t))_j\right|, \tag{2.50}$$

however,

$$|(H(q(t))Gx(t))_i| \le \sum_{j=1}^{g} |h_{ij}(q(t))||(Gx(t))_j|, \tag{2.51}$$

noticing that $H(q(t))$ and ω have non-negative components, then,

$$\begin{aligned}|(H(q(t))Gx(t))_i| &\le \sum_{j=1}^{g} h_{ij}(q(t))|(Gx(t))_j| \\ &\le \sum_{j=1}^{g}\left\{h_{ij}(q(t))\ \omega_j\ \frac{|(Gx)_j|}{\omega_j}\right\} \\ &\le \sum_{j=1}^{g}\{h_{ij}(q(t))\omega_j\}\ \underset{j}{Max}\left\{\frac{|(Gx)_j|}{\omega_j}\right\} \\ &\le (H(q(t))\omega)_i\ \underset{i}{Max}\left\{\frac{|(Gx)_i|}{\omega_i}\right\}.\end{aligned}$$

Substituting into (2.50)

$$\Delta V(x(t)) \le \underset{i}{Max}\left\{\left(\frac{(H(q(t))\omega)_i}{\omega_i} - 1\right)\underset{i}{Max}\left\{\frac{|(Gx(t))_i|}{\omega_i}\right\}\right\},$$

which is equivalent to

$$\Delta V(x(t)) \leq \left(\frac{(H(q(t))\omega)_i}{\omega_i} - 1\right) \underset{i}{Max} \left\{\frac{|(Gx(t))_i|}{\omega_i}\right\}.$$

Therefore, according to (2.48), the rate of increase $\Delta V(x(t))$ is strictly negative, then the function $V(x(t))$ is a common Lyapunov function for all stationary configurations of the uncertain system (2.35) for every $q \in \Gamma$, and consequently, the system is asymptotically stable.

For the continuous-time case, the non-quadratic function given by (2.49) is positive definite. Let us compute its derivative along the trajectories of the system (2.35)

$$\begin{aligned}\dot{V}(x(t)) &= \lim_{\varepsilon \to 0^+} \frac{V(x(t) + \varepsilon \dot{x}(t)) - V(x(t))}{\varepsilon} \\ &= \lim_{\varepsilon \to 0^+} \frac{1}{\varepsilon} \left\{ \underset{i}{Max} \left\{ \frac{|(Gx(t) + \varepsilon G(A + BF)x(t))_i|}{\omega_i} \right\} - V(x(t)) \right\}.\end{aligned}$$

According to (2.46):

$$\dot{V}(x(t)) = \lim_{\varepsilon \to 0^+} \frac{1}{\varepsilon} \left\{ \underset{i}{Max} \left\{ \frac{|((1 + \varepsilon H)Gx(t))_i|}{\omega_i} \right\} - V(x(t)) \right\},$$

this expression is equivalent to (2.50) for the discrete-time case where matrix H is replaced by $1 + \varepsilon H$, and consequently, the same reasoning leads to,

$$\dot{V}(x(t)) \leq \lim_{\varepsilon \to 0^+} \frac{1}{\varepsilon} (\underset{i}{Max} \left\{ \left(\frac{(|1 + \varepsilon H(q(t))|\omega)_i}{\omega_i} - 1 \right) V(x(t)) \right\}).$$

However, since matrix $H \in \mathscr{M}_H$ and as $\varepsilon \to 0$:

$$\begin{aligned}|1 + \varepsilon H(q(t))| &= \begin{cases} \varepsilon |h_{ij}| = \varepsilon h_{ij} & \text{for } i \neq j, \\ 1 + \varepsilon h_{ij} & \text{for } i = j, \end{cases} \\ &= 1 + \varepsilon H(q(t)).\end{aligned}$$

Replacing in 2.52, one obtains:

$$\dot{V}(x(t)) \leq \underset{i}{Max} \left\{ \left(\frac{(H(q(t))\omega)_i}{\omega_i} \right) V(x(t)) \right\}. \tag{2.52}$$

Therefore, according to (2.48), $\dot{V}(x(t))$ is strictly negative definite, and then, the function $V(x(t))$ is a common Lyapunov function for all stationary configuration of the uncertain system (2.35) for every $q \in \Gamma$, and consequently, the system is asymptotically stable. □

The re-formulation of all conditions enabling to solve the RLCRP, in both cases of continuous-time and discrete-time systems, as linear inequalities is now achieved. It is clear now that solving the RLCRP is a feasibility problem. Further, an objective function ε can be introduced as follows:

$$\text{for the discrete-time case, } H_{ij}\ \omega < \varepsilon\omega,\ 0 < \varepsilon \leq 1,$$
$$\text{for the continuous-time case, } H_{ij}\ \omega < -\varepsilon\omega,\ \varepsilon \geq 0,$$

for $i = 1, ..., \mu_A$, $j = 1, ..., \mu_B$. The choice of such objective function plays a double role. In fact, it is directly connected to the assigned spectrum in closed-loop [6, 22], and to the rate of increase in the Lyapunov function (see demonstration of Proposition 2.2 above). Hence, optimizing ε implies that the rate of convergence of the state to the origin is augmented. Solution of the linear programs proposed hereafter allows to find matrices H_{ij} for $i = 1, ..., \mu_A$, $j = 1, ..., \mu_B$ together with matrix T. Finally, it is obvious that the gain feedback F solution to the RLCRP is easily deduced from this solution.

Comments 2.1 :
(1) It is worth noting here that solving the RLCRP without linear programming is a complicated problem and needs a number of trial and error with an important computation burden to obtain the adequate regulator [6]. Using linear programming gives nice solution and does not need the trial and error procedure as proposed in previous works. But, it is not, in the other hand, easy to satisfy all required conditions. The fact is looking for a robust static regulator respecting constraints is not a simple problem.
(2) We note also, here, that matrices H_{ij} are used instead of working with the $H_{ij}^+ = sup(H_{ij}, 0)$ and $H_{ij}^- = sup(-H_{ij}, 0)$ as several other works [7, 19, 23], which are interested in the same kind of linear programming formulation. The problem in our sense is that there is a fundamental dependence between matrices H_{ij}^+ and H_{ij}^- which is omitted. It comes out that taking into account this fundamental dependence, that is the product $H_{ij}^+ H_{ij}^-$ must be zero since they are representing the $+$ and $-$ part of the same matrix H_{ij}, makes the optimization problem nonlinear. Hence, the linear programming approach can no more be sufficient to solve the problem.

2.4 Examples

Discrete-time case: Consider the linear uncertain discrete-time system given by (2.34) with:

$$A(q_A(t)) = \begin{bmatrix} 0.785 + q_1 & 0.5 \\ -0.4 & 1.2 \end{bmatrix}, \qquad B(q_B(t)) = \begin{bmatrix} 0.6 \\ -0.1 + q_2 \end{bmatrix}$$

The vertices of the domain of uncertainties are: $\nu_{11} = (0.015, 0)$, $\nu_{12} = (0.015, -0.02)$, $\nu_{21} = (-0.015, 0)$, $\nu_{22} = (-0.015, -0.02)$. Matrix G and vector ρ are given by:

$$G = \begin{bmatrix} -2.24 & 1.2 \\ -2.912 & 1.56 \end{bmatrix}, \omega = \begin{bmatrix} 30 \\ 20 \end{bmatrix}$$

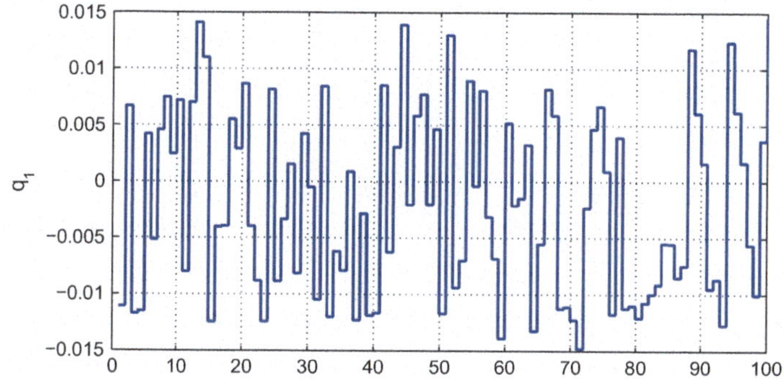

Fig. 2.1 q_1 evolution

and $\rho = 22$, $rank(G) = 1$.

We impose fictitious suitable constraints, in order to have $rank(G) = 2$, as follows:

$$G_a = \begin{bmatrix} -2.24 & 1.2 \\ -2.912 & 1.56 \\ 1.14 & -1.96 \end{bmatrix}, \ \omega_a = \begin{bmatrix} 30 \\ 20 \\ 20 \end{bmatrix}$$

The solution obtained by solving the proposed linear program are $\varepsilon = 0.977$, and

$$H_{11} = \begin{bmatrix} 0.6411 & 0.0000 & 0.2135 \\ 0 & 0.6411 & 0.2775 \\ 0 & 0.0000 & 0.9187 \end{bmatrix}, \ H_{12} = \begin{bmatrix} 0.6295 & 0 & 0.2061 \\ 0.0653 & 0.5792 & 0.2679 \\ 0 & 0.0147 & 0.9308 \end{bmatrix}$$

$$H_{21} = \begin{bmatrix} 0.5976 & 0.0000 & 0.1868 \\ 0.0832 & 0.5335 & 0.2428 \\ 0 & 0.0171 & 0.9322 \end{bmatrix}, \ H_{22} = \begin{bmatrix} 0.5859 & 0.0000 & 0.1794 \\ 0.1484 & 0.4717 & 0.2332 \\ 0.0000 & 0.0317 & 0.9443 \end{bmatrix}$$

and

$$T = \begin{bmatrix} 0.3257 & 0.1234 & 0.3083 \end{bmatrix}$$

leading to the following robust controller:

$$F = \begin{bmatrix} -0.7372 & -0.0211 \end{bmatrix}.$$

Evolution of q_1 (resp. q_2) is given in Fig. 2.1 (resp. Fig. 2.2). Figure 2.3 shows the positive invariance and the admissibility of the domain $\mathscr{D}(G, \omega)$ as well as the asymptotic stability of the system motion, from the different initial state $x_o = [80 \ \ 150]^T$, $x_o = [100 \ \ 100]^T$, and $x_o = [60 \ \ 50]^T$. However, Fig. 2.4 shows the control admissibility for $t > 0$.

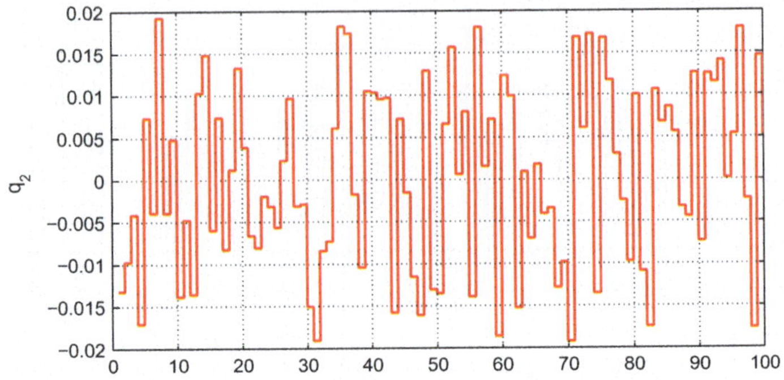

Fig. 2.2 q_2 evolution

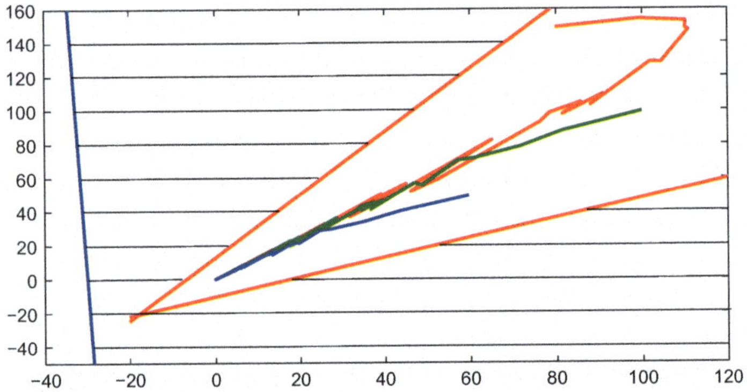

Fig. 2.3 States evolution

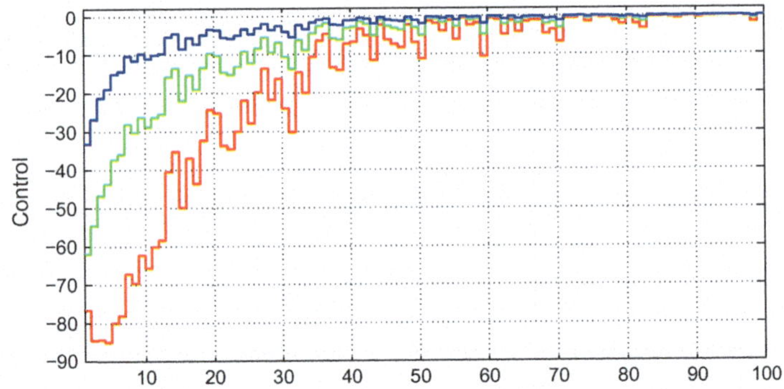

Fig. 2.4 Control evolution

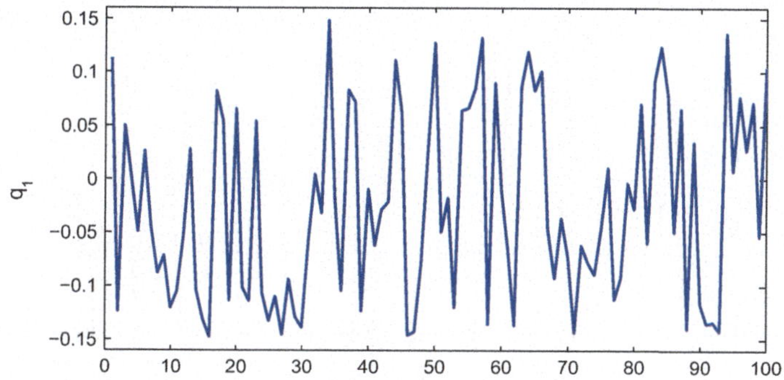

Fig. 2.5 q_1 evolution

Continuous-time case: Consider the linear uncertain continuous-time system given by (2.34) where,

$$A(q_A(t)) = \begin{bmatrix} -1.35 + q1 & 0.5 \\ -3.5 + q2 & 2.25 \end{bmatrix}, \quad B = \begin{bmatrix} 0.15 \\ 0.7 \end{bmatrix}$$

$$G = \begin{bmatrix} 2.5 & -1 \\ 1 & -2 \end{bmatrix}, \quad \omega = \begin{bmatrix} 20 \\ 30 \end{bmatrix}$$

and $\rho = 80$.

The vertices of the domain of uncertainties are:

$$\nu_1 = (-0.15, -0.1), \quad \nu_2 = (-0.15, 0.1),$$

$$\nu_3 = (0.15, -0.1) \text{ and } \nu_4 = (0.15, 0.1).$$

The solution obtained by the proposed linear program are: $\varepsilon = 0.301$ and

$$H_1 = \begin{bmatrix} -1.7685 & 0.0063 \\ 28.9849 & -0.1115 \end{bmatrix}, \quad H_2 = \begin{bmatrix} -1.9325 & 0.0001 \\ 27.5739 & -0.1475 \end{bmatrix},$$

$$H_3 = \begin{bmatrix} -1.7486 & 0.0035 \\ 28.9837 & -0.1214 \end{bmatrix}, \quad H_4 = \begin{bmatrix} -1.9325 & 0.0001 \\ 27.5739 & -0.1475 \end{bmatrix},$$

and $T = \begin{bmatrix} 1.1000 & 1.4500 \end{bmatrix}$. These results lead to the robust controller

$$F = \begin{bmatrix} 4.2000 & -4.0000 \end{bmatrix}.$$

Evolution of q_1 (resp. q_2) is given by Fig. 2.5 (resp. Fig. 2.6). Fig. 2.7 shows the positive invariance and the admissibility of the domain $\mathscr{D}(G, \omega)$ as well as the asymptotic stability of the system motion, from different initial states $x_o = [10 \ \ 20]^T$, $x_o = [-20 \ \ 20]^T$, and $x_o = [-25 \ \ -20]^T$. However, the Fig. 2.8 shows the control admissibility for all $t > 0$.

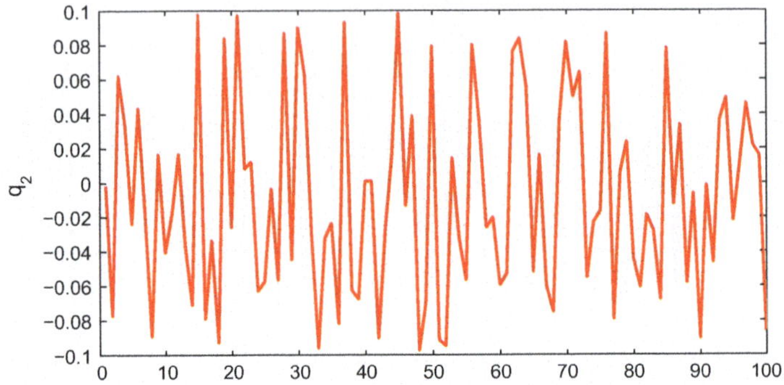

Fig. 2.6 q_2 evolution

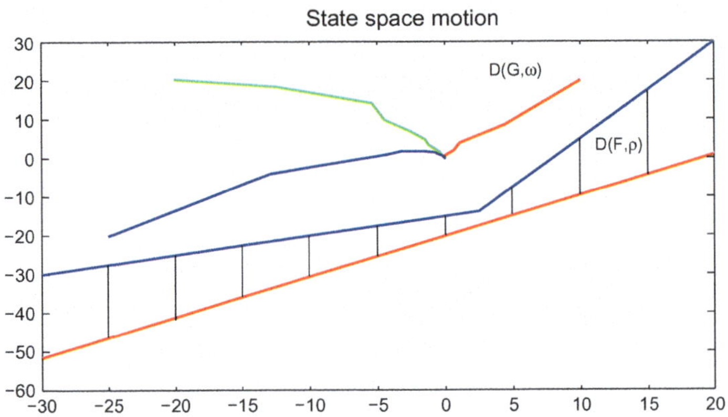

Fig. 2.7 States evolution

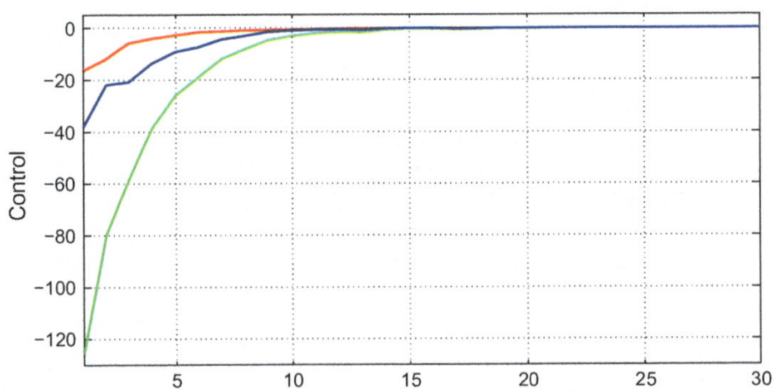

Fig. 2.8 Control evolution

2.5 Conclusion

In this chapter, we have generalized the solution to the LCRP for both continuous-time and discrete-time systems. The extension, when the system parameters are not perfectly known, is obtained in order to study robustness of such regulators. The solution is established using the positive invariance and the admissibility criteria. Hence, the asymptotic stability without constraints state and control violation is guaranteed. The obtained conditions are re-formulated using the Haar's lemma under matrix inequalities. These inequalities enable us to present the solution as linear programming algorithms. Illustrative examples are given in both of continuous-time case and discrete-time cases to show the ease of such approach to hold a complicated problem.

Further, to solve the problem, the weakest assumption for the rank of matrix G is considered, and the problem of dependence between matrices H^+ and H^- is overcome.

It has been shown that the obtained controllers can also be designed by selecting the desired closed-loop performance, followed by computing the solution of an algebraic equation for the nominal system, and finally checking the conditions of robustness at the vertices of the convex set of perturbations.

References

1. J.C. Doyle, K. Glover, P.P. Khargonekar, B.A. Francis, State space solution to H_2 and H_∞ control problems. IEEE Trans. Aut. Control. **34**(N° 8), 831–847 (1989)
2. G. Feng, M. Palaniswami, Y. Zhu, Stability of rate constrained robust pole placement adaptive control systems. Syst. Control. Lett. **18**, 99–107 (1992)
3. B.A. Francis, G. Zames, Design of H_∞ optimal multivariable feedback systems in *Proceedings of the CDC*, (1983), pp. 103–108
4. V.L. Kharitonov, Asymptotic stability of an equilibrium position of a family of systems of linear differential equations. Diffrential nye Uravnenya **14**, 2086–2088 (1978)
5. A.L. Zenlentsovsky, Non-quadratic Lyapunov functions for robust stability analysis of linear uncertain systems. IEEE Trans. Aut. Control. **39**(N° 1) (1994)
6. A. Benzaouia, F. Mesquine, Regulator problem for uncertain linear discrete-time systems with constrained control. Int. J. Robust Nonlinear Control. **4**, 387–395 (1994)
7. G. Bitsoris, E. Garvalou, Robust linear controller under state and control constraints in *Proceedings of the 31th CDC*, Tucson, Arizona (1992), pp. 2640–2642
8. A. Benzaouia, A. Hmamed, Regulator problem for linear continuous-time systems with non-symmetrical constrained control. IEEE Trans. Aut. Control. **38**(N° 10), 1556-1560 (1993)
9. A. Benzaouia, C. Burgat, Regulator problem for discrete-time systems with non-symmetrical constrained control. Int. J. Control, Vol. 48, N° 6(2441–2451) 1988
10. F. Blanchini, Feedback control for linear time-invariant systems with state and control bounds in the presence of disturbances. IEEE Trans. Aut. Control. **35**, 1231–1234 (1990)
11. F. Blanchini, Set Invariance in control. Automatica **35**, 1747–1767 (1999)
12. M.A. Dahleh, I.J. Diaz-Bobillo, *Control of Uncertain Systems: A Linear Programming Approach* (Prentice Hall, Engelwoods Cliffs, NJ, 1995)
13. Z. Lin, A. Saberi, Semi global exponential stabilization of linear discrete-time systems subject to input saturation via linear feedbacks. Syst. Control. Lett. **21**, 225–239 (1995)

14. F. Mesquine, F. Tadeo, A. Benzaouia, Regulator problem for linear systems with constraints on control and its increment or rate. Automatica **40**(N° 8), 1387-1395 (2004)
15. V.F. Montagner, P.L.D. Peres, S. Tarbouriech, I. Queinnec, Improved estimation of stability regions for uncertain linear systems with saturating actuators: an LMI-based approach in *Proceedings of the 45th CDC*, San diego, CA, USA (2006)
16. F. Mesquine, A. Benlamkadem, F. Tadeo, Robust constrained control for continuous time systems: an application to a ph process in *IEEE CCA conference*, Glasgow, Scotland, U.K., (2002), pp. 391–396
17. F. Mesquine, A. Benlemkadem, A. Benzaouia, Robust constrained regulator problem for linear uncertain systems. J. Dyn. Control. syst. **10**(N° 4) 527-544 (2004)
18. L. Benvenuti, L. Farina, Constrained control for uncertain discrete-time linear systems. Int. J. Robust Non-linear Control. **8**, 555–565 (1998)
19. B.E.A. Milani, A.N. Carvalho, Robust linear regulator design for discrete-time systems under polyhedral constraints. Automatica **31**, 1489–1493 (1995)
20. M. Vassilaki, G. Bitsoris, Constrained regulation of linear continuous-time dynamical systems. Syst. Control. Lett. **13**, 247–252 (1989)
21. M. Vassilaki, J.C. Hennet, G. Bitsoris, Feedback control of linear discrete-time system under state and control constraints. Int. J. Control. **47**, 1727–1735 (1988)
22. A. Benzaouia, F. Mesquine, M. Naib, A. Hmamed, Robust pole assignment in complex plane regions for linear uncertain constrained control systems. Int. J. Syst. Sci. **32**(N° 1), 83–89 (2001)
23. J.C. Hennet, E.B. Castelan, Robust invariant controllers for constrained linear systems in *ACC*, (1992), pp. 993–997

Chapter 3
Constrained Control and Rate or Increment for Linear Systems

3.1 Introduction

In real or physical plants, as seen before, the most frequent constraints are of saturation type: limitations on the magnitude of certain quantities or variables. One could cite a main approach to study such problem: the positive invariance concept ([2–4, 7, 33]). Apart from magnitude saturation constraints, this chapter deals with a different type of constraints, namely rate or incremental constraints. These constraints were introduced while considering predictive control and practical applications [1, 9] and other approaches [16, 30].

Recall that for this class of linear dynamical systems, having both constraints on control magnitude and rate or increment, Lin in [20] showed that all asymptotically null controllable bounded input systems are semi-globally stabilizable through linear feedback in the presence of both constraints. A method to stabilize a particular plant in the presence of constraints on both input magnitude and increments was considered by [31]. Other approaches have been presented, for example, by [10–12, 14, 17, 19–21, 32].

Henceforth, this chapter investigates the problem of locally stabilizing linear continuous-time and discrete-time systems with limitations on both control magnitude and control increment or rate.

The positive invariance approach, used here, gives a simple solution to stability and performance of stabilizable dynamical linear systems with bounded input and increment or rate. Necessary and sufficient conditions of positive invariance for incremental domains with respect to autonomous systems are given. Further, a link is done between pole assignment procedure and these conditions to design stabilizing controllers by state feedback, respecting control constraints and control increment or rate constraints also.

Furthermore, those results are extended to the case of non-symmetrical constrained systems with additive bounded disturbances. Necessary and sufficient conditions of positive invariance for incremental domains with respect to (w.r.t.) autonomous systems with additive bounded disturbances are then derived.

A. Benzaouia et al., *Saturated Control of Linear Systems*, Studies in Systems, Decision and Control 124, https://doi.org/10.1007/978-3-319-65990-9_3

It is worth noticing here that the perturbation rejection as studied thereafter was also studied in [7, 8] using the set invariance concept and in [18] using convex writing of the saturation function. But, there were no constraints on the rate of the control.

Hence, this chapter is the extension to those works using the positive invariance concept to deal with rate and magnitude control constraints as a first part and in the presence of additive bounded disturbances in a second part.

3.2 Regulator Problem for Linear Systems with Constraints on Control and Its Increment or Rate

3.2.1 Problem Statement

Consider a linear time invariant system represented in the state space by:

$$\delta x(\cdot) = Ax(\cdot) + Bu(\cdot) \tag{3.1}$$

where $x(\cdot) \in \mathbb{R}^n$ is the state of the system, $u(\cdot) \in \mathbb{R}^m$ is the input vector, which is constrained to evolve in the following domain:

$$\Omega = \{u \in \mathbb{R}^m / \quad -u_{\min} \leq u \leq u_{\max}, \quad u_{\min}, u_{\max} \in \mathbb{R}_+^{m*}\}. \tag{3.2}$$

(i) For discrete-time systems, the control increment is constrained as follows:

$$-\Delta_{\min} \leq u(k+1) - u(k) \leq \Delta_{\max} \tag{3.3}$$

(ii) For continuous-time systems, the control rate is constrained as follows:

$$-\Delta_{\min} \leq \dot{u}(t) \leq \Delta_{\max} \tag{3.4}$$

We denote:

$$U = \begin{bmatrix} u_{\max} \\ u_{\min} \end{bmatrix}, \quad \Delta = \begin{bmatrix} \Delta_{\max} \\ \Delta_{\min} \end{bmatrix}.$$

The problem studied in this first part is the following.

Find a stabilizing linear state feedback as:

$$u(\cdot) = Fx(\cdot), \quad F \in \mathbb{R}^{mxn} \tag{3.5}$$

such that the closed-loop system is asymptotically stable respecting constraints on both the control signal magnitude and the control increment (or, in the continuous-time case, derivative).

Consider the linear time-invariant autonomous system:

$$\delta z(\cdot) = Hz(\cdot), \quad z(t_o) = z_o, \tag{3.6}$$

where $z \in \mathbb{R}^m$ is the state, constrained to evolve in the domain:

$$\mathscr{D}_z = \{z \in \mathbb{R}^m / -z_{\min} \leq z \leq z_{\max}\ z_{\min}, z_{\max} \in \mathbb{R}_+^{m*}\}, \tag{3.7}$$

- For discrete-time systems by

$$-\Delta_{\min} \leq z(k+1) - z(k) \leq \Delta_{\max} \tag{3.8}$$

- For continuous-time systems by

$$-\Delta_{\min} \leq \dot{z}(t) \leq \Delta_{\max} \tag{3.9}$$

To design a state feedback control that might ensure constraints fulfillment for both control and rate or increment, we begin first by establishing conditions such that state rate or increment constraints for the autonomous system (3.6) are respected. Further, for the proposed state feedback control presented later, control rate dynamics are separated from control dynamics and hence can be studied sequentially. It will be easy then, to mix the conditions obtained separately for both control and increment or rate constraints to obtain the controller that respects both of them. The following lemma studies the fulfillment of incremental or stare rate constraints for motion of system (3.6).

Lemma 3.1 *The evolution of the autonomous system (3.6) respects incremental or rate constraints if and only if matrix H satisfies:*

$$\widetilde{(H - \mathbb{I})}_d\ \vartheta \leq\ \Delta \quad \textit{for the discrete-time case} \tag{3.10}$$

$$\widetilde{H}_d\ \vartheta\ \leq \Delta \qquad \textit{for the continuous-time case.} \tag{3.11}$$

where

$$\vartheta = \begin{bmatrix} z_{\max} \\ z_{\min} \end{bmatrix}, \quad \Delta = \begin{bmatrix} \Delta_{\max} \\ \Delta_{\min} \end{bmatrix}.$$

Proof <u>*If part*</u>: Consider the discrete-time case and assume that condition (3.10) is satisfied.

Hence, it is possible to write:

$$\begin{aligned} z(k+1) - z(k) &= Hz(k) - z(k) \\ &= (H - \mathbb{I})z(k) \\ &= G\ z(k), \end{aligned}$$

where we noted $G := H - \mathbb{I}$, but it is known that:

$$-z_{\min} \leq z(k) \leq z_{\max}. \tag{3.12}$$

Next, decompose matrix $G = G^+ - G^-$, pre-multiplying (3.12) by G^+ and $-G^-$, gives:

$$-G^+ z_{\min} \leq G^+ z(k) \leq G^+ z_{\max}, \tag{3.13}$$

$$-G^- z_{\max} \leq -G^- z(k) \leq G^- z_{\min}, \tag{3.14}$$

the addition of inequalities (3.13) and (3.14) enables us to write:

$$-G^+ z_{\min} - G^- z_{\max} \leq G z(k) \leq G^+ z_{\max} + G^- z_{\min},$$

according to condition (3.10):

$$-\Delta_{\min} \leq -G^+ z_{\min} - G^- z_{\max} \leq G z(k) \leq G^+ z_{\max} + G^- z_{\min} \leq \Delta_{\max},$$

which is equivalent to:

$$-\Delta_{\min} \leq z(k+1) - z(k) \leq \Delta_{\max}.$$

In the continuous-time case:

$$\dot{z}(t) = H\, z(t)$$

following the same reasoning, replacing matrix G by matrix H and condition (3.10) by condition (3.11), it is easy to obtain:

$$-\Delta_{\min} \leq \dot{z}(t) \leq \Delta_{\max}$$

Only if part: Now, consider the continuous-time case: Assume that the derivative of $z(t)$ respects the constraints, and that condition (3.11) is not satisfied for an index $1 \leq i \leq n$ such that

$$[\widetilde{H}_d\ \vartheta]_i > \Delta_i \tag{3.15}$$

that is

$$[H^+ z_{\max} + H^- z_{\min}]_i > \Delta^i_{\max}$$

Then, the following state vector for the system can be selected:

$$\phi(t) = \begin{cases} z^j_{\max} & \text{if } \ h_{ij} > 0 \\ 0 \ \text{ if } \ h_{ij} = 0 \\ -z^j_{\min} & \text{if } \ h_{ij} < 0 \end{cases}, \quad j = 1, \ldots, n$$

It is easy to check that $\phi(t)$ is an admissible state for the system. Calculation of the ith component of the derivative of this state gives:

$$\begin{aligned}[\tfrac{d}{dt}\phi(t)]_i &= [H\phi(t)]_i = \sum_{j=1}^{n} h_{ij}\phi_j(t) \\ &= [H^+ z_{\max} + H^- z_{\min}]_i\end{aligned}$$

taking into account inequality (3.15), it is possible to write:

$$[\frac{d}{dt}\phi(t)]_i > \Delta^i_{\max}$$

which contradicts the assumption. The discrete-time case part could be easily deduced replacing matrix H by matrix G in the necessary part. □

By virtue of Lemma 3.1, if conditions (3.10) or (3.11) are satisfied, the evolution of the autonomous system (3.6) respects the incremental or state rate constraints. But it was assumed that the state $z(t)$ does not leave domain $\mathscr{D}_z$ given by (3.7) which is not guaranteed in general case. To do so, one has to ensure, in addition, the positive invariance of the former domain. Hence, the evolution of the autonomous system (3.6) will respect both constraints on the state $z(t)$ and constraints on its increment or rate.

Positive invariance conditions of polyhedral domains of type $\mathscr{D}_z$ have already been reported by [2] and by [3]. Combining these conditions to those proposed in Lemma 3.1 enables us to claim the following:

Lemma 3.2 *Domain (3.7) is positively invariant with respect to motion of system (3.6), and incremental or rate constraints (3.3) or (3.4) are respected if and only if:*

$$\begin{cases} (i) \ \ \widetilde{(H - \mathbb{I})}_d \ \vartheta \ \leq \ \Delta \\ (ii) \ \ \widetilde{H}_d \ \vartheta \ \leq \ \vartheta \end{cases} \textit{for the discrete-time case.} \tag{3.16}$$

$$\begin{cases} (i) \ \ \widetilde{H}_d \ \vartheta \ \leq \ \Delta \\ (ii) \ \ \widetilde{H}_c \ \vartheta \ \leq \ 0 \end{cases} \textit{for the continuous-time case.} \tag{3.17}$$

Proof First, notice that the dynamics of the system state and those of the state increment or rate are independent. Second, conditions (3.16-ii) and (3.17-ii) imply that the magnitude bounds hold, and sequentially, conditions (3.16-i) and (3.17-i) imply that the rate or increment bounds hold also. □

Relating the previous lemma to the pole assignment procedure presented in the first chapter is the cornerstone to solve the problem as stated above.

3.2.2 Regulator with Constraints on Control and Its Rate or Increment

Consider a stabilizable linear time invariant system (3.1) with constraints on both control magnitude (3.2) and control increments or rate (3.3) or (3.4). Using the state feedback:

$$u(\cdot) = Fx(\cdot), \; F \in \mathbb{R}^{mxn} \tag{3.18}$$

such that

$$\sigma(A + BF) \in \mathbb{D} \tag{3.19}$$

The following domain of linear behavior is induced in the state space:

$$\mathscr{D}_F = \{x \in \mathbb{R}^n / \quad -u_{\min} \le Fx \le u_{\max}, \quad u_{\min}, u_{\max} \in \mathbb{R}_+^{m*}\} \tag{3.20}$$

If the state does not leave the domain (3.20), the control magnitude signal does not violate the constraints. That is, the set $\mathscr{D}_F$ is positively invariant with respect to motion of system (3.1). This gives the following result:

Proposition 3.1 *System (3.1) with state feedback (3.18)–(3.19) is asymptotically stable at the origin with constraints on both the control magnitude and its increment or rate if there exists a matrix $H \in \mathbb{R}^{mxm}$ satisfying conditions (1.37)–(1.40), such that:*

(i)

$$F\ A + F\ B\ F = H\ F \tag{3.21}$$

(iia) for the discrete-time case:

$$\widetilde{(H - \mathbb{I})}_d\ U \le \Delta \tag{3.22}$$

$$\widetilde{H}_d\ U \le U \tag{3.23}$$

(iib) for the continuous-time case:

$$\widetilde{H}_d\ U \le \Delta \tag{3.24}$$

$$\widetilde{H}_c\ U \le 0 \tag{3.25}$$

where $U = [u_{\max}^T \quad u_{\min}^T]^T$ for all initial state $x_o \in \mathscr{D}_F$.

Proof Introduce the following change of coordinates: $z = F\ x$, it is possible to write

$$\begin{aligned}\delta z(\cdot) &= F\ \delta x(\cdot) = F\ (\ A + B\ F\ )\ x(\cdot)\\ &= H\ F\ x(\cdot) = H\ z(\cdot).\end{aligned} \tag{3.26}$$

With this transformation, domain $\mathscr{D}_F$ is transformed to domain $\mathscr{D}_z$ given by (3.7). Further, with conditions (3.23) and (3.25), it is easy to note that domain $\mathscr{D}_z$ is positively invariant with respect to the system (3.26), while the constraints on the increment of the control are respected. Bearing in mind that $\sigma(A + B F) \in \mathbb{D}$ and that the linear behavior is guaranteed, one can conclude to the asymptotic stability of the closed-loop system. □

Remarks 3.1 • It is worth noting that conditions (3.23) and (3.25) do not affect the set of positive invariance $\mathscr{D}_F$. However, they present an additional constraint on the pole assignment problem.
• Observe that the eigenvalues of matrix H can be used to impose closed-loop eigenvalues, so it gives a useful tuning parameter to select the desired feedback performance.

Algorithm 3.1
1. Check if matrix A has $(n - m)$ stable eigenvalues, if not, augment the matrix B with $(n - m)$ null columns.
2. Choose matrix $H \in \mathbb{R}^{m\times m}$ or $H \in \mathbb{R}^{n\times n}$, if the system is augmented, according to (1.37)–(1.40), (3.22) and (3.23) or (1.37)–(1.40), (3.24) and (3.25). (see comments below on how to choose matrix H).
3. Compute the gain matrix F or F_a by using (1.42).
4. Use F or extracted F from F_a for the control.

Comments:
(1) At step 1 of Algorithm 3.1, when augmenting the system is necessary, the fictitious input limitations are first chosen and then the procedure for selecting matrix H, as proposed below, is achieved. This choice is made by trial and error and may appear restrictive. But, in our sense, it is a further degree of freedom that can be exploited. In fact, these constraints can be chosen as soft as to enlarge the domain $\mathscr{D}_F$ as needed.
(2) It is worth noting that larger is the domain $\mathscr{D}_F$, slower are the dynamics assigned in closed loop from matrix H [5]. For the so-called inverse procedure and the algorithm above, there is an important step of choosing the matrix H satisfying all required conditions and assigning the closed-loop spectrum. Important cases are presented below.
(2.i) When the goal is to enlarge the initial conditions set, linear programming can be used to select the suitable matrix H.
• For the discrete-time case, the following linear program can be used:

$$\text{minimize } \varepsilon$$
$$\text{subject to}$$

$$\begin{bmatrix} H^+ & H^- \\ H^- & H^+ \end{bmatrix} \begin{bmatrix} u_{max} \\ u_{min} \end{bmatrix} \leq \varepsilon \begin{bmatrix} \phi_1 \\ \phi_2 \end{bmatrix} \tag{3.27}$$

$$0 < \varepsilon \leq 1, \quad H^+ \geq 0, \quad H^- \geq 0 \tag{3.28}$$

where $\phi_1 < u_{max}$ and $\phi_2 < u_{min}$ are tuning vectors. In fact, if the feasible solution H^+, H^- is such that the matrix $H = H^+ - H^-$ satisfies required conditions (1.37)–(1.40) and (3.22), use matrix H in the Algorithm 3.1. Else, change vectors ϕ_1 and ϕ_2 according to rules hereafter.

Rules of changing vectors ϕ_1 and ϕ_2 can be divided into two main cases: First, the linear programming in discrete systems case has a feasible solution that does not satisfy (1.37)–(1.40), in this case, a small perturbation of components of vectors ϕ_1 and/or ϕ_2 leads to sensible changes of eigenvalues and eigenvectors of matrix H such that the required condition can be fulfilled. Second, given feasible solution of linear programming in discrete systems case that satisfy (1.37)–(1.40), we note that condition (3.22) is not fulfilled. For unfeasible lines in this last condition, respective component of vectors ϕ_1 and/or ϕ_2 should be changed by trial and error to remove unfeasibility.

Similar linear program with the goal of finding both matrices H and F can be found in [15]. But, in our sense, the necessity of matrix B of the system to be square and non-singular, in order to make the optimization problem linear, can be very restrictive.

- For the continuous-time case, the following linear program can be used:

$$\text{maximize } \varepsilon$$
$$\text{subject to}$$

$$\begin{bmatrix} H^+ & H^- \\ H^- & H^+ \end{bmatrix} \begin{bmatrix} u_{max} \\ u_{min} \end{bmatrix} \leq \begin{bmatrix} \Delta_{max} \\ \Delta_{min} \end{bmatrix} \tag{3.29}$$

$$\begin{bmatrix} H_1 & H_2 \\ H_2 & H_1 \end{bmatrix} \begin{bmatrix} u_{max} \\ u_{min} \end{bmatrix} \leq -\varepsilon \begin{bmatrix} \phi_1 \\ \phi_2 \end{bmatrix} \tag{3.30}$$

$$\varepsilon \geq 0, \quad H^+ \geq 0, \quad H^- \geq 0 \tag{3.31}$$

where vectors $\phi_1 \geq 0$ and $\phi_2 \geq 0$ are tuning vectors. If the feasible solution H^+, H^- is such that matrix $H = H^+ - H^-$ satisfies condition (1.37)–(1.40) use matrix H for the algorithm. Else, change ϕ_1 and ϕ_2 according to rules hereafter.
In this case, rules of changing vectors ϕ_1 and ϕ_2 consist of a small perturbation of components of vectors ϕ_1 and/or ϕ_2. Sensible changes of eigenvalues and eigenvectors of matrix H are then obtained such that the required condition can be fulfilled.
(2.ii) If the initial condition set is imposed or given, linear programming can also be used to select matrix H. Two possibilities then exist: First, to use linear programming in discrete-time systems case or linear programming in continuous-time systems case where inequalities defining the initial condition set are added as constraints. Second, to use a linear program that determines both matrices H and the state feedback F [24, 28].

(2.iii) If the goal is to robustly assign the closed-loop spectrum in a specified complex plane region, algorithm given in [6] is appropriate to be used to select the suitable matrix H.

Example 3.1 Continuous-time case
Consider the double integrator in the continuous-time state space representation given by [13]:

$$A = \begin{bmatrix} 0 & 1 \\ 0 & 0 \end{bmatrix}, \quad B = \begin{bmatrix} 0 \\ 1 \end{bmatrix}$$

Control constraints are as follows:

$$-1 \leq u(t) \leq 5$$

Assume that control rate is constrained as follows:

$$-10 \leq \dot{u}(t) \leq 3$$

As discussed in the remark above, the system can be augmented with fictitious constrained inputs v in domain $[-v_{min} \ \ v_{max}]$. First, the fictitious constraints are selected as follows:

$$-2 \leq v(t) \leq 5; \quad -5 \leq \dot{v}(t) \leq 2$$

Processing the linear program presented above for the continuous-time case enables to select the suitable matrix H as:

$$H = \begin{bmatrix} -1.483 & 0.043 \\ 0 & -1 \end{bmatrix}$$

which satisfies all the required conditions (1.37)–(1.40), with $\varepsilon = 1$, that is

$$\begin{aligned} \widetilde{H}_c U &= [-7.19 \ \ -5 \ \ -1.39 \ \ -2]^T \leq 0, \\ \widetilde{H}_d U &= [1.69 \ \ 2 \ \ 7.5 \ \ 5]^T \\ &\leq [3 \ \ 2 \ \ 10 \ \ 5]^T. \end{aligned}$$

Solution of Eq. (3.21) leads to the following augmented gain matrix F_a:

$$F_a = \begin{bmatrix} -1.48 & -2.48 \\ 34.23 & 23.08 \end{bmatrix}.$$

Note that the effective gain matrix F is extracted from the first row of the previous matrix (the second row corresponds to the fictitious input). The effective gain matrix is then:

$$F = \begin{bmatrix} -1.48 & -2.48 \end{bmatrix}$$

The closed-loop dynamics are given by:

$$A + BF = \begin{bmatrix} 0 & 1 \\ -1.48 & -2.48 \end{bmatrix}.$$

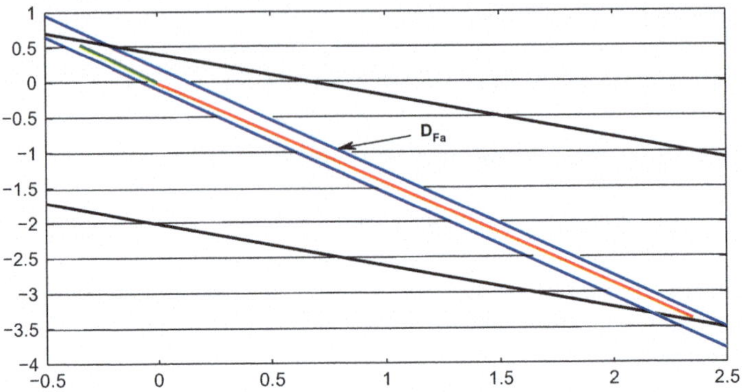

Fig. 3.1 Domain of linear behavior $\mathscr{D}_{F_a}$

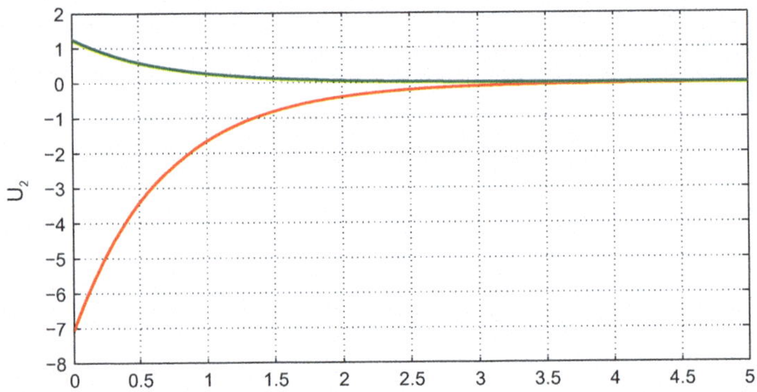

Fig. 3.2 Control evolution in time

As it has been pointed out before, the eigenvalues of matrix H can be used as tuning parameters to impose closed-loop dynamics: It is easy to note here that $\sigma(A + BF) = \sigma(H)$.

The obtained set of positive invariance with the augmented matrix F_a is given by:

$$\mathscr{D}_{F_a} = \left\{x \in \mathbb{R}^n \mid -w_{min} \leq F_a x \leq w_{max}\right\}$$

where $w_{min}^T = [u_{min}^T \; v_{\min}^T]$ and $w_{max}^T = [u_{max}^T \; v_{\max}^T]$.

Figure 3.1 represents the set of linear behavior $\mathscr{D}_{F_a}$, and in red, a trajectory emanating from the initial state $x_o = [+2.3514 \;\; -3.364]^T$ and in green from $x_o = [-0.3445 \;\; 0.5249]^T$ lying in the set $\mathscr{D}_{F_a}$. Figure 3.2 represents the admissible control evolution for the two initial conditions with the same colors. Finally, Fig. 3.3 shows the evolution of the control rate. It is possible to see that the states evolution respects the constraints.

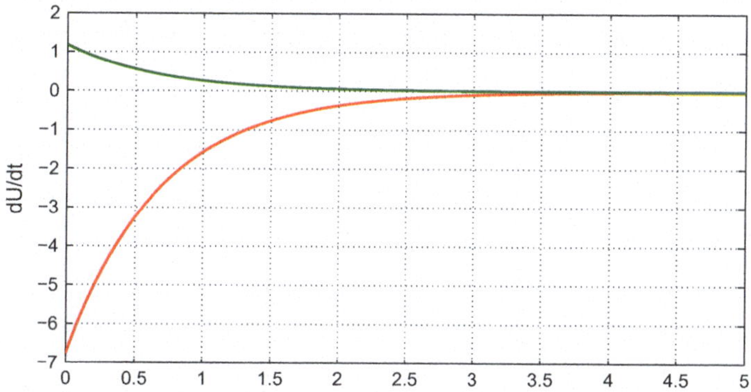

Fig. 3.3 Control's rate evolution in time

Example 3.2 Discrete-time case
Let us consider the system given by:

$$A = \begin{bmatrix} 1.785 & 0.5 \\ 1.4 & 1.2 \end{bmatrix}, \quad B = \begin{bmatrix} 0.6 \\ -0.1 \end{bmatrix}.$$

The constraints are given as follows:

$$-2 \leq u(k) \leq 5, \qquad -3 \leq u(k+1) - u(k) \leq 7$$
$$U = \begin{bmatrix} 5 \\ 2 \end{bmatrix} \text{ and } \Delta = \begin{bmatrix} 7 \\ 3 \end{bmatrix}.$$

Note that $\sigma(A) = \{2.3788; 0.6062\}$, hence the system admits 1 stable eigenvalue that will be kept in closed loop. The unstable one will be replaced. Let us choose matrix H, in this case, it is a scalar $H = 0.8$ and it is the eigenvalue to be assigned in closed loop. The requested conditions may be checked as follows:

$$\widetilde{(H - \mathbb{I})}_d U - \Delta = \begin{bmatrix} -8 \\ -3.4 \end{bmatrix} \leq 0,$$
$$\tilde{H}_d U - U = \begin{bmatrix} -3 \\ -5.4 \end{bmatrix} \leq 0.$$

The eigenvector of the stable eigenvalue is given by:

$$\xi_1 = \begin{bmatrix} -0.2206 \\ 1.3201 \end{bmatrix},$$

one can check that $B\xi_1 \neq 0$. The state feedback will be then given by:

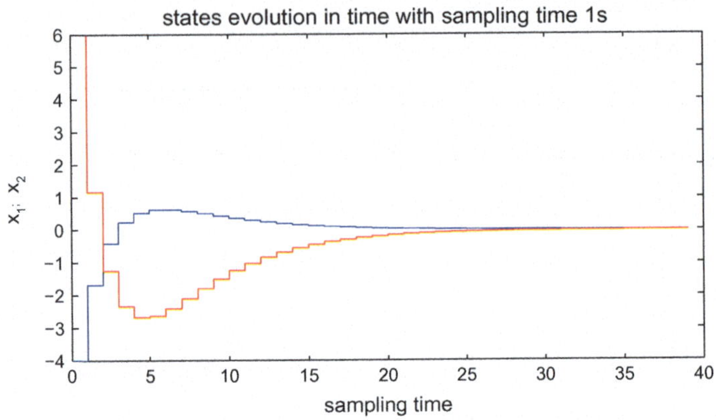

Fig. 3.4 State evolution

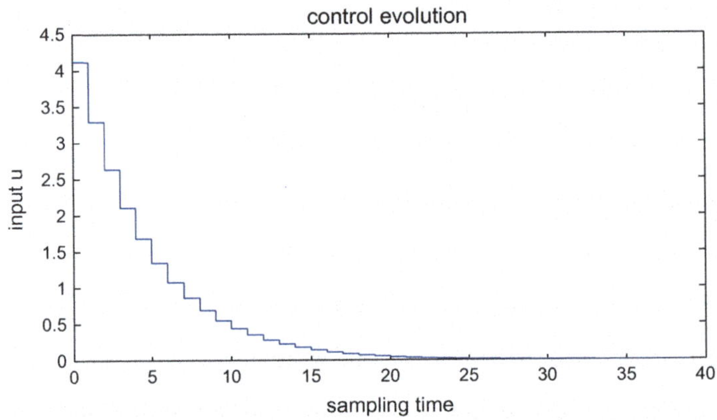

Fig. 3.5 Control evolution

$$F = \begin{bmatrix} 1 & 0 \end{bmatrix} \times \begin{bmatrix} \xi_1 & \xi_2 \end{bmatrix}^{-1} = \begin{bmatrix} -2.8315 & -1.2010 \end{bmatrix},$$

where $\xi_2 = (\lambda \mathbb{I} - A)^{-1} \times B$ (in this case $\lambda = H$ and the associated eigenvector is 1. The closed-loop dynamics are given by:

$$\sigma(A + B\ F) = \{\, 0.6\ ,\quad 0.8\, \}.$$

It is worth noting here that the domain of admissible states is not bounded in this case. See Figs. 3.4, 3.5, and 3.6.

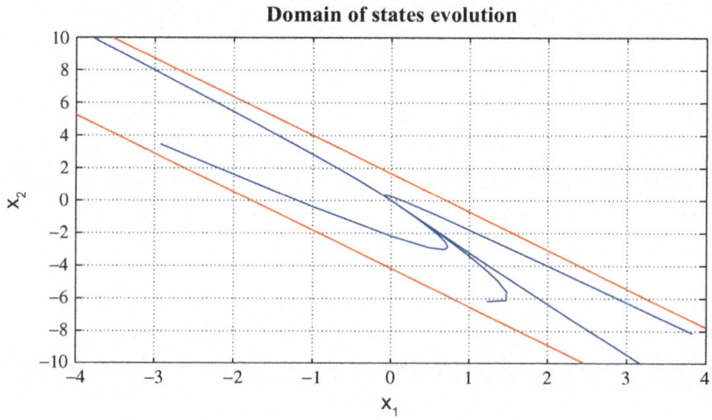

Fig. 3.6 Positive invariance domain

3.3 Constrained Control and Rate or Increment for Linear Systems with Additive Disturbances

After solving above, the problem of stabilization of linear systems with constraints on control and increment or rate, the extension to the same case of systems subject to bounded disturbances, as in [26, 27], is studied.

3.3.1 *Problem Statement*

Consider the linear time invariant system:

$$\delta x(t) = A\ x(t)\ +\ B\ u(t)\ +\ E\ p(t) \tag{3.32}$$

where $x(t) \in \mathbb{R}^n$ is the state of the system, $u(t) \in \mathbb{R}^m$ is the input, constrained to evolve in the domain given by (3.2). The control rate or increment is constrained as:

$$\begin{aligned} -\Delta_{\min} \quad &\leq \quad \dot{u}(t) \quad \leq \quad \Delta_{\max} \text{ for continuous-time case} \\ -\Delta_{\min} \leq \ & u(t+1) - u(t) \ \leq \Delta_{\max} \text{ for discrete-time case} \end{aligned} \tag{3.33}$$

$p(t)$ is an additive disturbance bounded as:

$$-\ p_{min} \leq p(t) \leq p_{max} \tag{3.34}$$

Further, we denote:

$$U = \begin{bmatrix} u_{\max} \\ u_{\min} \end{bmatrix}, \quad \Delta = \begin{bmatrix} \Delta_{\max} \\ \Delta_{\min} \end{bmatrix}, \quad \varpi = \begin{bmatrix} p_{\max} \\ p_{\min} \end{bmatrix}$$

The problem, to be solved, is to find stabilizing linear state feedback as (3.5), ensuring closed-loop asymptotic stability of the system despite perturbations with no violation of non-symmetrical constraints on the rate (or increment) and magnitude of the control.

3.3.2 Controller Design with Constraints on Control and Rate with Disturbances

Let us now extend the previous results presented in the first part of this chapter to the case of systems with additive bounded disturbances. To do so, a technical lemma that was established in [23] is revisited in the case of bounded disturbances. Relating that lemma to the pole assignment procedure presented earlier enables to find stabilizing controllers for systems with non-symmetrical constrained control and rate or increment. Consider the linear time invariant autonomous system:

$$\delta z(\cdot) = Hz(\cdot) + Ep(\cdot), \quad z(t_o) = z_o, \tag{3.35}$$

where $z \in \mathbb{R}^m$ is the state.

Lemma 3.3 *The evolution of the autonomous system (3.35), with additive disturbances $p(\cdot)$, respects rate or increment constraints if and only if matrix H satisfies*

$$\widetilde{H}_d \ \vartheta \ + \widetilde{E}_d \ \varpi \ \leq \Delta \quad \textit{for the continuous-time case} \tag{3.36}$$

$$\widetilde{(H \ - \mathbb{I})}_d \ \vartheta \ + \widetilde{E}_d \ \varpi \ \leq \Delta \quad \textit{for the discrete-time case} \tag{3.37}$$

where

$$\vartheta = \begin{bmatrix} z_{\max} \\ z_{\min} \end{bmatrix}, \quad \Delta = \begin{bmatrix} \Delta_{\max} \\ \Delta_{\min} \end{bmatrix}, \quad \varpi = \begin{bmatrix} p_{\max} \\ p_{\min} \end{bmatrix}.$$

Proof Let us first begin by the proof for the continuous-time case.
<u>*If part*</u>: Assume that condition (3.36) is satisfied, then

$$- z_{\min} \leq z(t) \leq z_{\max}, \tag{3.38}$$

next, decompose matrix $H = H^+ - H^-$; pre-multiplying (3.38) by H^+ and $-H^-$, gives:

$$- H^+ z_{\min} \leq H^+ z(t) \leq H^+ z_{\max} \tag{3.39}$$

$$- H^- z_{\max} \leq -H^- z(t) \leq H^- z_{\min} \tag{3.40}$$

Further, consider the bounds on the disturbance $p(t)$:

$$-p_{\min} \leq p(t) \leq p_{\max} \tag{3.41}$$

Applying to (3.41) the same technique used to achieve (3.39) and (3.40) where matrix $E = E^+ - E^-$ leads to the following inequalities:

$$-E^+ p_{\min} \leq E^+ p(t) \leq E^+ p_{\max} \tag{3.42}$$

$$-E^- p_{\max} \leq -E^- p(t) \leq E^- p_{\min} \tag{3.43}$$

Addition of the inequalities (3.39)–(3.40) and (3.42)–(3.43) enables to write the following:

$$\begin{aligned} -E^+ p_{\min} - E^- p_{\max} - H^+ z_{\min} - H^- z_{\max} &\leq Hz(t) + Ep(t) \\ &\leq H^+ z_{\max} + H^- z_{\min} + E^+ p_{\max} + E^- p_{\min} \end{aligned}$$

according to condition (3.36),

$$-\Delta_{\min} \leq -G^+ z_{\min} - G^- z_{\max} \leq Gz(t) \leq G^+ z_{\max} + G^- z_{\min} \leq \Delta_{\max}$$

this is equivalent to:

$$-\Delta_{\min} \leq \dot{z}(t) \leq \Delta_{\max}.$$

Only if part: Now, we assume that the derivative of $z(t)$ respects the constraints, and we suppose that condition (3.36) was not satisfied for an index $1 \leq i \leq n$ such that:

$$[\widetilde{H}_d \vartheta]_i + [\widetilde{E}_d \varpi]_i > \Delta_i \tag{3.44}$$

expanding (3.44):

$$[H^+ z_{\max} + H^- z_{\min}]_i + [E^+ p_{\max} + E^- p_{\min}]_i > \Delta^i_{\max}$$

Then, the following state vector for the system can be selected:

$$\phi(t) = \begin{cases} z^j_{\max} \text{ if } h_{ij} > 0 \\ 0 \text{ if } h_{ij} = 0 \\ -z^j_{\min} \text{ if } h_{ij} < 0 \end{cases}, j = 1, ..., n$$

It is easy to check that $\phi(t)$ is an admissible state for the system. Further, the following admissible perturbation may occur:

$$\kappa(t) = \begin{cases} p^j_{\max} \text{ if } e_{ij} > 0 \\ 0 \text{ if } e_{ij} = 0 \\ -p^j_{\min} \text{ if } e_{ij} < 0 \end{cases}, \ j = 1, ..., p$$

Calculation of the ith component of the derivative of this state gives

$$\begin{aligned} [\tfrac{d}{dt}\phi(t)]_i &= [H\,\phi(t) \ + E\,p(t)]_i \\ &= \sum_{j=1}^{n} h_{ij}\phi_j(t) + \sum_{j=1}^{p} e_{ij}\kappa_j(t) \\ &= [H^+ z_{\max} + H^- z_{\min} + E^+ p_{max} + E^- p_{min}]_i \end{aligned}$$

taking into account inequality (3.44), it is possible to write:

$$[\frac{d}{dt}\phi(t)]_i > \Delta^i_{\max}$$

which contradicts the assumption that the derivative respects the constraints.

For the discrete-time case, assume that condition (3.37) holds true and writes the increment as:

$$\begin{aligned} z(k+1) - z(k) &= Hz(k) + Ep(k) - z(k) \\ &= (H - \mathbb{I})z(k) + Ep(k) \\ &= Gz(k) + Ep(k) \end{aligned}$$

making the same reasoning as the continuous-time case with matrix H replaced by G, the proof of the if part may be easily deduced.

The necessary part is also deduced replacing matrix H by matrix G and the derivative of the state $\phi(t)$ by the increment $\phi(t+1) - \phi(t)$. □

With this background, we are now able to solve the problem stated in Sect. 3.2: consider the linear time invariant stabilizable system with additive disturbances and constraints on both control and rate of the controls (3.32)–(3.34). Using the state feedback:

$$u(t) = F\,x(t), \ F \in \mathbb{R}^{m\times n}, \ \sigma(A + B\,F) \in \boldsymbol{D} \tag{3.45}$$

induces the domain (3.20) of linear behavior in the state space.

If the state does not leave the domain (3.20), the control signal does not violate the constraints. That is, the set $\mathscr{D}_F$ is $\mathscr{D}_P$-positively invariant with respect to motion of system (3.32). This gives the following result:

Proposition 3.2 *The disturbed system (3.32) with state feedback (3.45) is asymptotically stable at the origin from all initial states $x_o \in \mathscr{D}_F$, respecting constraints on both the control and its rate (or increment), if there exists a matrix $H \in \mathbb{R}^{m\times m}$ satisfying conditions (1.37)–(1.40), such that:*

$$(i)\ \ F\,A + F\,B\,F = H\,F \tag{3.46}$$

$$(iia)\ \ \begin{cases} \widetilde{H}_d\,U + \widetilde{(F\,E)}_d\,\varpi \le \Delta \\ \widetilde{H}_c\,U + \widetilde{(F\,E)}_d\,\varpi \le 0 \end{cases} \text{for the continuous-time case.} \tag{3.47}$$

$$(iib)\ \ \begin{cases} \widetilde{(H - \mathbb{I})}_d\,U + \widetilde{(F\,E)}_d\,\varpi \le \Delta \\ \widetilde{H}_d\,U + \widetilde{(F\,E)}_d\,\varpi \le U \end{cases} \text{for the discrete-time case.} \tag{3.48}$$

Proof Introducing the following change of coordinates $z = F\,x$, it is possible to write (3.32) as follows:

$$\begin{aligned} \delta z(t) &= F\delta x(t) = F(A + B\,F)\,x(t) + F\,E\,p(t) \\ &= H\,F\,x(t) + F\,Ep(t) = H\,z(t) + F\,E\,p(t) \end{aligned} \tag{3.49}$$

With this transformation, domain $\mathscr{D}_F$ is transformed into domain $\mathscr{D}_z$ given by (3.7). Further, with conditions (3.47), it is easy to note that domain $\mathscr{D}_z$ is $\mathscr{D}_P$-positively invariant with respect to the system (3.49), while the constraints on the increment of the control are respected. Bearing in mind that $\sigma(A + BF) \in \mathbb{D}$ and that the linear behavior is guaranteed, it is possible to conclude the asymptotic stability of the closed-loop system. □

The steps to follow for design of such controllers are the same as those proposed in the Algorithm 3.1 presented above and reported in [25] for systems with non-additive disturbances. Methods to choose the matrix H for different cases were widely discussed in [25].

Comments 3.1 It is true that the set of positive invariance in this case is not the absolute maximal but it is maximal with respect to the chosen feedback F. In fact, for a given matrix feedback F, the maximal set where no violation of control constraints may occur is the set $\mathscr{D}_F$ as proposed above. Nevertheless, piecewise techniques [5] or maximization procedure for the set of positive invariance [25] can be used to overcome this difficulty.

Some closed-loop specifications lead to pole assignment in a specified complex plane region. In this case, the spectrum of matrix H lies within a specified complex plane if conditions given in [6] are fulfilled. These conditions can easily be added as conditions on the matrix H.

3.3.3 The Maximal Disturbance Set

As vector ϑ and matrix $\widetilde{H}_c$ for the open-loop continuous-time case or $\widetilde{(H - \mathbb{I})}_d$ have positive components, inequalities (3.36) or (3.37) can never be satisfied if the difference $(\Delta - \widetilde{E}_d\varpi)$ was negative. Hence, the interesting conclusion is that inequality

(3.36) or (3.37) permits to compute the maximum perturbation set that can be admissible with these rate constraints; that is, the set given by :

$$\mathscr{D}_P^{max} = \{p(t) \in \mathbb{R}^p, -p_{\min} \leq p(t) \leq p_{\max} / \widetilde{E}_d \ \varpi = \Delta\}. \tag{3.50}$$

Since matrix E and vector Δ are known from the statement of the problem, it can be concluded if such rate constraints requirement could be fulfilled or not, and if it is admissible or not.

As stated above, the maximal disturbance set such that asymptotic stability of the closed-loop system with no violation of constraints on both rate or increment and control can be estimated. Consider the system with additive disturbances

$$\dot{x}(t) \ = \ A\ x(t) \ + \ B\ u(t) \ + \ E\ p(t), \tag{3.51}$$

stabilized by state feedback (3.45). The maximal disturbance allowed set can be estimated as follows.

Corollary 3.1 *The maximal disturbance set such that closed-loop asymptotic stability can be ensured and rate (or increment) and magnitude control constraints are not violated is given by:*

$$\mathscr{D}_P^{max} \ = \ \left\{p(t) \in \mathbb{R}^p / - p_{min}^{max} \leq p(t) \leq p_{max}^{max}\right\}, \tag{3.52}$$

where vector ϖ^{max} *satisfies*

$$\begin{aligned}
\widetilde{(F\ E)}_d\ \varpi^{max} &= min(\Delta - \widetilde{H}_d U, -\widetilde{H}_c\ U) \ \ \textit{for the continuous-time case},\\
\widetilde{(F\ E)}_d\ \varpi^{max} &= min(\Delta - \widetilde{(H - \mathbb{I})}_d U, U - \widetilde{H}_d\ U) \ \textit{for the discrete-time case},
\end{aligned}$$

the minimum here is taken component-wise.

Proof Let us begin by the continuous-time case and assume that the objective is to stabilize the system by state feedback with $p(t)$ a perturbation vector with unknown limits. From condition (3.47), one can write as follows:

$$\begin{cases}
\widetilde{(F\ E)}_d\ \varpi \ \leq \ \Delta \ - \ \widetilde{H}_d\ U\\
\widetilde{(F\ E)}_d\ \varpi \ \leq \ - \widetilde{H}_c\ U
\end{cases}$$

hence, the maximal disturbance set that can be seen as the limit of fulfillment of the previous two conditions. Further, for any vector $T \in \mathbb{R}^p$ such that $T \leq \varpi^{max}$, it is easy to check that condition (3.47) is satisfied.
For the discrete-time case, the proof is similar. □

Example 3.3 Application of the previous results to control of DC motor for positioning of circuits boards in a mount robot presented in [22, 27] is considered. The Y-axis of the machine uses a ball screw transmission driven by a current controlled

DC motor. The rotation of the DC motor is converted into a translation motion by ball screw. A positioning table attached to the ball nut carries different loads. The process can be simplified as a two mass system:

$$\begin{aligned} J_m \ddot{\theta}_m + b_m \dot{\theta}_m &= K_m i_m - T_f - T_l \\ T_l &= k_t(\theta_m - x_l/p) \\ m_l \ddot{x}_l + b_l \dot{x}_l &= T_l/p \end{aligned}$$

where θ_m and x_l are the motor angle and table displacement, respectively, i_m is the motor current, T_l is the load torque due to the torsion of ball screw, T_f is the friction torque, p is the screw pitch, J_m is the motor inertia plus ball screw inertia, b_m, b_l, and K_m are, respectively, the damping coefficients and the constant torque, k_t is the stiffness, m_l is the equivalent mass of load, table, and nut [22]. Converting the motor angle position to linear position and rewriting the model in the state space gives:

$$\dot{x}(t) = Ax(t) + Bu(t) - Bp(t)$$

where

$$A = \begin{bmatrix} 0 & 1 \\ 0 & -9.67 \end{bmatrix}, \quad B = \begin{bmatrix} 0 \\ 7.35 \end{bmatrix}$$

and $x = [\ p\theta \quad p\dot{\theta}\ ]^T$, $u(t) = i_m(t)$ and $p(t)$ is the lumped disturbance of load torque, friction, and other external disturbances. The current input of the DC motor, its rate, and the perturbation are, respectively, constrained as follows:

$$-4 \le u(t) \le 4, \quad -80 \le \dot{u}(t) \le 80, \quad -1 \le p(t) \le 1.$$

Comments 3.2 The origin of the rate constraints is that the limit of the variation of the motor current $i_m(t)$ is taken as a peak to peak value at a sampling time. As the system possesses one stable eigenvalue, in this case, it is not necessary to augment the system. Further, for this system, the constraints are symmetric from the original example, this fact simplifies a number of conditions but we insist to present the theoretical non-symmetrical case for the seek of generality. Furthermore, the resulting closed loop cannot be unstable as it is our goal to stabilize the system. Hence, assuming the perturbation is limited within the given set is always true although it is a function of the state of the system as presented.

Matrix H reduces to a scalar in this case, let us select $h = -\alpha$, where $\alpha \neq 0$, is any positive number which satisfies all the required conditions (1.37)–(1.40):

$$\widetilde{H}_c U \ + \widetilde{(-FB)}_d \varpi \le 0$$

It is clear that α here may be any positive number. To show the importance of conditions including the disturbances, we begin by the case where such conditions for

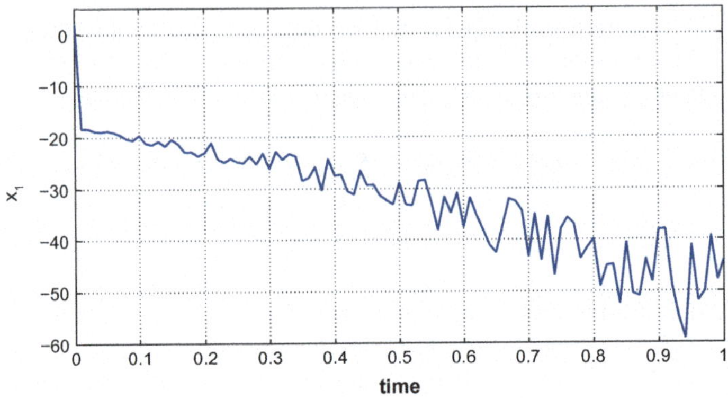

Fig. 3.7 State $x_1(t)$ motion

the disturbed system are not satisfied. Hence, let $h = -\alpha = -20$, then the stabilizing gain feedback is given by:

$$F = [-26.3129 \quad -2.7211]$$

It is easy to check that one of the two conditions is not satisfied, that is,

$$\widetilde{H}_d U + \widetilde{(-F B)}_d \varpi \leq \Delta$$

$$\begin{bmatrix} 4\alpha \\ 4\alpha \end{bmatrix} + \begin{bmatrix} 20 p_{min} \\ 20 p_{max} \end{bmatrix} > \begin{bmatrix} \Delta_{max} \\ \Delta_{min} \end{bmatrix}$$

Figure 3.7 shows that in fact the system motion, from the initial condition $x_o = [2 - 20]^T$, does not converge to the origin, especially state $x_1(t)$ is divergent (Fig. 3.7), even state $x_2(t)$ is convergent (Fig. 3.8). We show also the control evolution in Fig. 3.9 which does not respect the constraints. It must be pointed out that despite the initial state is chosen inside the set of linear behavior, the system motion does not converge to the origin. Figure 3.10 represents disturbance $p(t)$ variation.

Now choose $\alpha = 12$. Solution of Eq. (3.46) leads to the following gain matrix F:

$$F = [-15.7878 \quad -1.6327]$$

At this step, one has to check that all required conditions (3.47) are fulfilled. In fact:

$$\widetilde{H}_d U + \widetilde{(-F B)}_d \varpi \leq \Delta,$$

$$\begin{bmatrix} 4\alpha \\ 4\alpha \end{bmatrix} + \begin{bmatrix} 12 p_{min} \\ 12 p_{max} \end{bmatrix} \leq \begin{bmatrix} \Delta_{max} \\ \Delta_{min} \end{bmatrix},$$

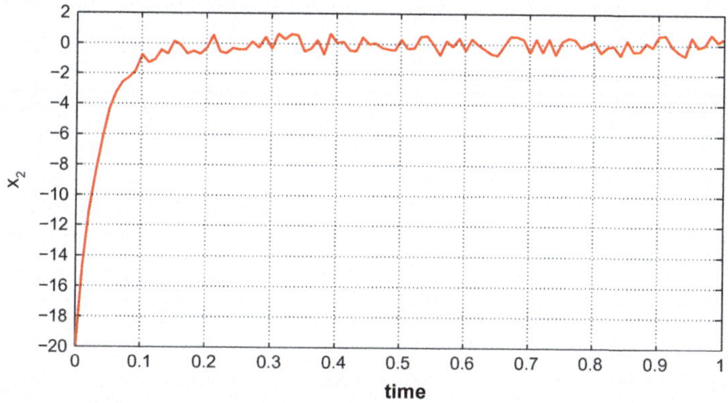

Fig. 3.8 State $x_2(t)$ motion

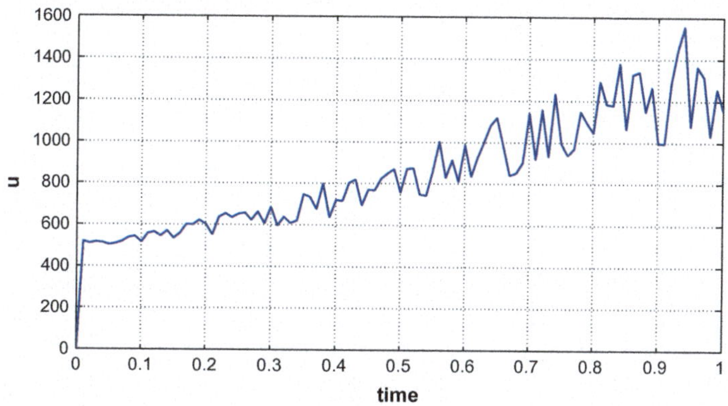

Fig. 3.9 Evolution of the control $u(t)$

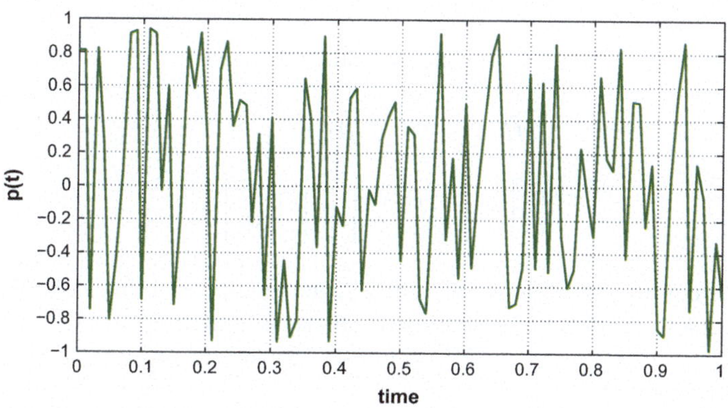

Fig. 3.10 Disturbance $p(t)$ evolution

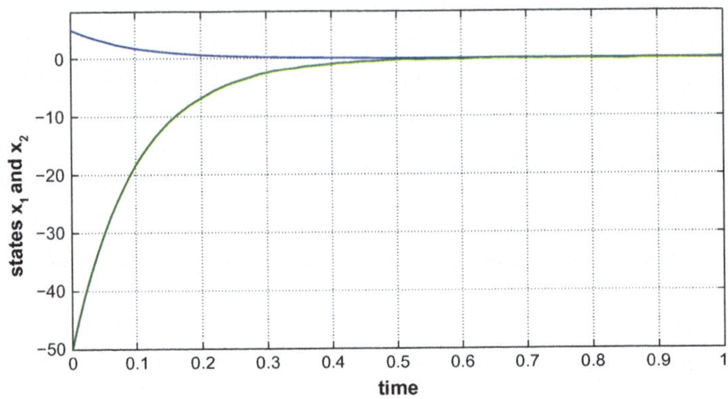

Fig. 3.11 State $x_1(t)$ and $x_2(t)$ motion

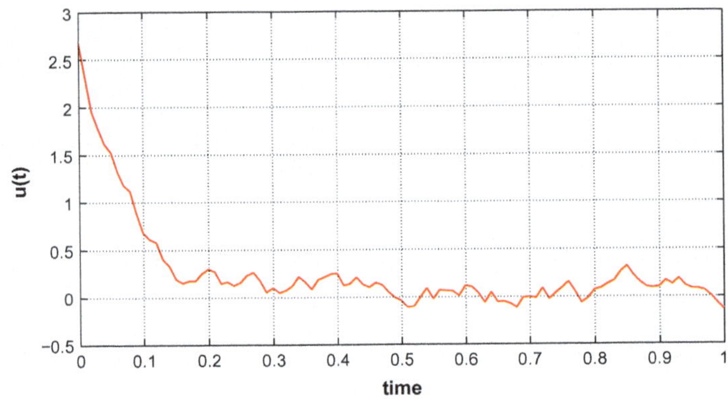

Fig. 3.12 Control evolution

that is, $4\alpha + 12p_{max} = 60 \leq 80$ which is satisfied. Then

$$\widetilde{H}_c U + (\widetilde{-F\,B})_d \varpi \leq 0,$$

$$\begin{bmatrix} -4\alpha \\ -4\alpha \end{bmatrix} + \begin{bmatrix} 12p_{max} \\ 12p_{min} \end{bmatrix} = \begin{bmatrix} -36 \\ -36 \end{bmatrix} \leq 0.$$

Finally, all required conditions are fulfilled, asymptotic stability is achieved with the obtained state feedback. Simulation results are summarized in Fig. 3.11 from the initial condition $x_o(t) = [5 \quad -50]^T$. Figures 3.12 and 3.13 plot input and rate's input evolution in time.

The maximal allowed disturbance set, which does not change stability and does not violate imposed constraints, in this case, can be easily deduced as follows:

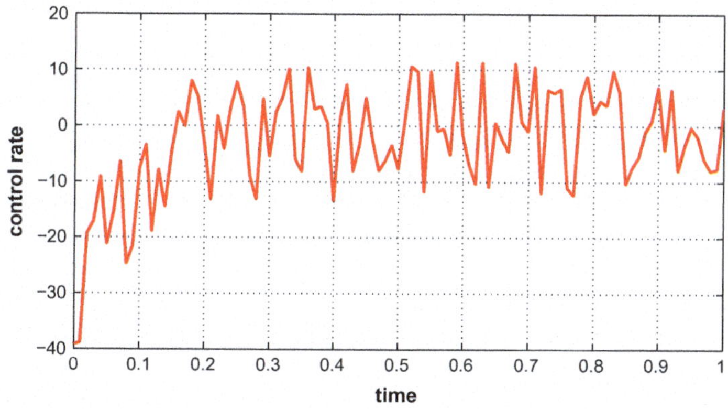

Fig. 3.13 Rate input evolution in time

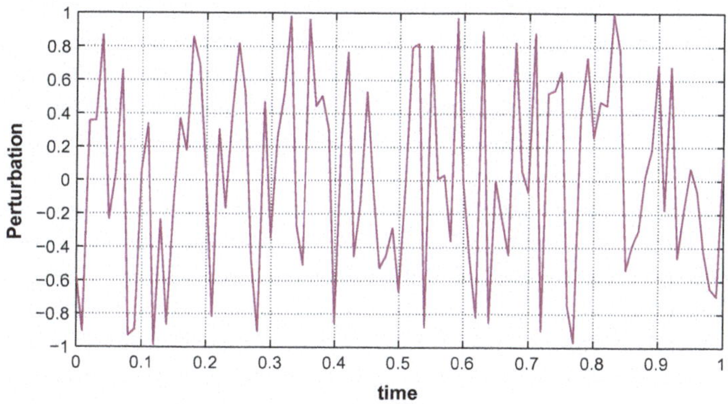

Fig. 3.14 Evolution of perturbation

$$(\widetilde{-F\ B})_d\ \varpi^{max} = min(\Delta,\ -\widetilde{H_c}U),$$

simple calculation leads to:

$$\varpi^{max} = \begin{bmatrix} 4 \\ 4 \end{bmatrix}$$

An example of the perturbation evolution in this case is given in Fig. 3.14.

3.4 Conclusion

In this chapter, the regulator problem for linear systems which contain independent input saturation and rate saturation nonlinearities has been studied. A method has been developed to avoid these constraints, maintaining the closed-loop system in a linear region. The method proposed in this chapter is based on determining a feedback control law that ensures positive invariance of a set included in a region of closed-loop linear behavior and including all admissible initial states. This positive invariant set is then considered as a linear local region of stability. The link to the so-called inverse procedure, the pole assignment method for constrained control, to the previous conditions is the cornerstone of this work. This link makes possible to develop a simple algorithm to compute a regulator respecting constraints on both control and its increment or rate. It must be pointed out that performance specifications in terms of the position of closed-loop eigenvalues can be easily incorporated in the design steps, which makes it a powerful technique. An example has been presented to show the application of the technique presented in this chapter to a double integrator plant: It has been seen how simple it is to apply this technique. Further, linear programming was used to select the suitable matrix H for design. Furthermore, work can be done to consider uncertainty in the plant and using linear matrix inequalities to solve the problem.

In an other hand, this chapter deals also with the same regulator problem for linear systems with non-symmetrical constrained control and rate or increment but in addition subject to additive bounded disturbances. Necessary and sufficient conditions, established for linear autonomous systems such that their motion respects rate constraints together with $\mathscr{D}_P$-positive invariance, are used in this case to solve this problem. The maximal disturbance set, such that robust asymptotic stability, rate and control constraints are not destroyed, has also been easily deduced.

The same problem can be solved for sampled-data control systems with magnitude and rate saturating control [29].

References

1. D. Agneli, A. Casavola, E. Mosca, Predictive PI-control of linear plants under positional and incremental input saturation. Automatica **36**(10), 1505–1516 (2000)
2. A. Benzaouia, C. Burgat, Regulator problem for linear discrete-time systems with non-symmetrical constrained control. Int. J. Control **48**(6), 2441–2451 (1988)
3. A. Benzaouia, A. Hmamed, Regulator problem of linear continuous-time systems with constrained control. IEEE Trans. Autom. Control **38**(10), 1556–1560 (1993)
4. A. Benzaouia, F. Mesquine, Regulator problem for uncertain linear discrete time systems with constrained control. Int J Robust Nonlinear Control **4**, 387–395 (1994)
5. A. Benzaouia, A. Baddou, Piecewise linear constrained control for continuous-time systems. IEEE Trans. Autom. Control **44**(7), 1477–1481 (1999)
6. A. Benzaouia, F. Mesquine, M. Naib, A. Hmamed, Robust pole assignment in complex plane regions for linear uncertain constrained control systems. Int. J. Syst. Sci. **32**(1), 83–89 (2001)

7. F. Blanchini, Feedback control for linear systems with state and control bounds in the presence of disturbance. IEEE Trans. Autom. Control **35**, 1131–1135 (1990)
8. F. Blanchini, Ultimate boundedness control for uncertain discrete time systems via set induced Lyapunov function. IEEE Trans. Autom. Control **AC-39**, 428–433 (1994)
9. D.W. Clarke, C. Mohtadi, P.S. Tuffs, Generalized predictive control, part 1: the basic algorithm; part 2: extension and interpretations. Automatica **23**(2), 137–160 (1987)
10. H. Fang, Z. Lin, T. Hu, Analysis of linear systems in the presence of actuator saturation and L_2 disturbances. Automatica **40**, 1229–1238 (2004)
11. G. Feng, M. Palaniswami, Y. Zhu, Stability of rate constrained robust pole placement adaptive control systems. Syst. Control Lett. **18**, 99–107 (1992)
12. E.G. Gilbert, K.T. Tan, Linear systems with state and control constraints: The theory and application of maximal output admissible sets. IEEE Trans. Autom. Control **36**, 1008–1020 (1991)
13. P.O. Gutman, P. Hagander, A new design of constrained controllers for linear systems. IEEE Trans. Autom. Control **30**, 22–33 (1985)
14. G.D. Hanson, R.F. Stengel, Effects of displacement and rate saturation on the control of statically unstable aircraft. AIAA J. Guidance Control Dyn. **7**, 197–205 (1984)
15. J.C. Hennet, J. P. Beziat, A class of invariant regulators for the discrete-time linear constrained regulation problem. Automatica **27**(3), 549–554 (1991)
16. D. Henrion, S. Tarbouriech, V. Kucera, Control of linear systems subject to input constraints: A polynomial approach. Automatica **36**(4), 597–604 (2001)
17. H. Hindi, S. Boyd, Analysis of linear systems with saturating control using convex optimization, in *Proceedings of the 37th IEEE CDC*. (Florida, USA, 1998)
18. T. Hu, Z. Lin, B.M. Chen, Analysis and design method for linear systems subject to actuator saturation and disturbance. Automatica **38**, 351–359 (2002)
19. P. Kapasouris, M. Athans, Control systems with rate and magnitude saturation for neutrally stable OL systems, in *Proceedings of the IEEE CDC* (Honolulu, Hawaii, U.S.A., 1990), pp. 3404–3409
20. Z. Lin, Semi-global stabilization of linear systems with position and rate-limited actuators. Syst. Control Lett. **30**, 1–11 (1997)
21. Z. Lin, M. Pachter, S. Banda, Y. Shamash, Stabilizing feedback design for linear systems with rate limited actuators, in *Control of Uncertain Systems with Bounded Inputs*. Lecture Notes in Control and Information Sciences, vol. 227 (Springer, 1997), pp. 173–186
22. Y.F. Li, J. Wikander, Discrete time sliding mode control of a DC motor and ball screw driven positioning table, in *Proceedings of 15th IFAC*, (Barcelona, Spain, 2002)
23. F. Mesquine, F. Tadeo, A. Benzaouia, Regulator problem for linear systems with constrained control and its increment, in *Proceedings of 15th IFAC*, (Barcelona Spain, 2002)
24. F. Mesquine, A. Benlemkadem, Robust linear constrained regulator problem. IMA J. Math. Control Inform. **24**, 81–94 (2007)
25. F. Mesquine, F. Tadeo, A. Benzaouia, Regulator problem for linear systems with constrained control and its increment or rate. Automatica **40**, 1387–1395 (2004)
26. F. Mesquine, F. Tadeo, A. Benzaouia, Constrained control and rate or increment for linear systems with additive disturbances. Math. Prob. Eng. (2006)
27. F. Mesquine, F. Tadeo, A. Benzaouia, Regulator constrained control and rate problem for linear systems with additive disturbances, in *Proceedings of american control conference*, (Boston, 2004)
28. B.E.A. Milani, A.N. Carvalho, Robust linear regulator design for discrete-time systems under polyhedral constraints. Automatica **31**, 1489–1493 (1995)
29. A.H.K. Palmeria, J.M. Gomes da Silva, S. Tarbouriech, I.M.F. Ghiggi, Stability analysis of sampled-data control systems under magnitude and rate actuators, in *European Control Conference*, (Linz, Austria, 2015)
30. A.A. Stoorvogel, A. Saberi, P. Sannuti, Performance with regulation constraints. Automatica **36**, 1443–1456 (2000)

31. L. Trygve, R. Murray, T.Y. Fossen, Stabilization of integrator chains in the presence of magnitude and rate saturation; a gain scheduling approach, in *Proceedings of IEEE CDC*, (San Diego, USA, 1997), pp. 4004–4005
32. F. Tyan, D.S. Bernstein, Dynamic output feedback compensation for linear systems with independent amplitude and rate saturation. Int. J. Control **67**(1), 89–116 (1997)
33. M. Vassilaki, J.C. Hennet , G. Bitsoris, Feedback control of linear discrete-time systems under state and control constraints. Int. J. Control **47**(6), 1727–1735 (1988)

Chapter 4
Regulator Problem for Singular Linear Systems with Constrained Control

4.1 Introduction

This chapter studies the stabilization of singular linear systems with input saturation and constraints on the control rate. Since singular linear systems can model many systems in electrical circuit networks, robotic, and economics. Singular systems have been of great interest in the control literature, [10, 17, 19, 24]. Hence, main classical concepts and results obtained for linear systems have already been extended to singular systems [6, 7, 11–13, 16, 20, 26]. The constraints considered in this chapter are asymmetric (the maximum and the minimum value of the constraints do not have the same absolute value) [3, 5, 21, 25], because constraints in real systems are usually asymmetric. As a regulator for this kind of systems, this chapter proposes state-feedback PD controllers, where state and derivative feedback are used to obtain more flexibility of design and performance than those obtained using only proportional state-feedback.

Many works have discussed the design of state and derivative feedback controllers for linear systems (see [8, 9, 14, 23] and the references therein). The main task of derivative feedback gain F_2 is to transform the singular system into a non-singular one [18]. However, this derivative feedback can produce a matrix $E + BF_2$ which is ill-conditioned. Generally, the use of equivalent representation of singular systems, obtained with a state augmentation can lead to a synthesis of the PD controller with no need of inversion of matrix $E + BF_2$.

Thus, this chapter presents necessary and sufficient conditions of positive invariance for singular systems with constraints on the control and its rate by using a PD controller. As a particular case, results for PD controllers for linear systems with constraints on the control and its rate are obtained.

A. Benzaouia et al., *Saturated Control of Linear Systems*, Studies in Systems, Decision and Control 124, https://doi.org/10.1007/978-3-319-65990-9_4

4.2 Regulation of Singular Linear System Under Constrained Control Magnitude and Rate

4.2.1 Problem Formulation

The problem which is the regulation of singular linear system with constrained control magnitude and rate is formulated below. Consider the following system described by:

$$E\dot{x}(t) = Ax(t) + Bsat(u(t)) \qquad (4.1)$$
$$x(0) = x_o$$

where $x \in \mathbb{R}^n$ is the state, $u \in \mathbb{R}^m$ is the control, and $Rank(E) = r \leq n$. The control and its rate are assumed here to be constrained as follows:

$$u \in \Omega = \{u \in \mathbb{R}^m | -u_{min} \leq u \leq u_{max};\ u_{max}, u_{min} \geq 0\} \qquad (4.2)$$
$$\dot{u} \in \Omega_\Delta = \{u \in \mathbb{R}^m | -\Delta_{min} \leq \dot{u} \leq \Delta_{max};\ \Delta_{max}, \Delta_{min} \geq 0\} \qquad (4.3)$$

Then, vectors U and Δ are defined as follows:

$$U = \begin{bmatrix} u_{max} \\ u_{min} \end{bmatrix} \in \mathbb{R}^{2m},\ \Delta = \begin{bmatrix} \Delta_{max} \\ \Delta_{min} \end{bmatrix} \in \mathbb{R}^{2m}.$$

Throughout the chapter, it is assumed that:

(**AS1**) (E,A,B) is PD stabilizable: that is, $Rank[sE - AB] = n,\ \forall s \in \mathbb{C}$ such that $Re(s) > 0$.

(**AS2**) The state x and its derivative $\dot{x}$ are available for measure.

For this chapter, some useful definitions and preliminary results can be reviewed from Chap. 1 on singular systems. The objective of this work is to design for the singular linear system (4.1) a stabilizing controller of the form:

$$u(t) = F_1x(t) - F_2\dot{x}(t), \qquad (4.4)$$

(state and state derivative feedback) that respects constraints on the control and its rate. To reach a solution, it is only necessary to select F_2 such that $G_1 = (E + BF_2)^{-1} \in \mathbb{R}^{n\times n}$ exists. Then, the unsaturated system (4.1) can be written as:

$$\dot{x}(t) = G_1Ax(t) + G_1BF_1u(t) \qquad (4.5)$$
$$x(0) = x_o$$

Then, the unsaturated system in closed loop is described by:

$$\dot{x}(t) = G_1(A + BF_1)x(t) \qquad (4.6)$$
$$= A_rx(t),$$

and the control can be expressed as follows:

$$\begin{aligned} u(t) &= (F_1 - F_2 A_r)x(t) \\ &= (F_1 - F_2 G_1 A - F_2 G_1 B F_1)x(t) = Kx(t), \end{aligned} \tag{4.7}$$

with $K \in \mathbb{R}^{m\times n}$. With this system in closed loop, the induced set of constraints on the state is given by:

$$\mathscr{D} = \left\{x \in \mathbb{R}^n | - u_{min} \leq Kx \leq u_{max}\right\}. \tag{4.8}$$

Note that from Eq. (4.7) one can write:

$$K = MF_1 - F_2 G_1 A \tag{4.9}$$
$$M = \mathbb{I}_m - F_2 G_1 B. \tag{4.10}$$

The main idea of the approach of positive invariance is to impose that the system trajectories evolve only inside the region of linear behavior defined by (4.8), so that Eq. (4.6) remains valid.

The proposed representation of the singular system is used to design the PD controller in the presence of constraints on both the control and its rate. The obtained regular system enables to deal easily with the problem inside the set of linear behavior $\mathscr{D}$ for the synthesis of the proposed stabilizing controller. That is the synthesis of the gain matrices F_1 and F_2 [5].

4.2.2 *Positive Invariance for Singular Linear Systems*

Necessary and sufficient condition for the positive invariance of the set $\mathscr{D}$ given by (4.8) with respect to system (4.6) is presented by using a constrained PD controller. Consider the dynamic of the control obtained with Eqs. (4.7), (4.9)–(4.10), hence, $\dot{u}(t) = KA_r x(t)$. If there exists a matrix $H \in \mathbb{R}^{m\times m}$ such that $KA_r = HK$, then the control dynamic is:

$$\dot{u}(t) = Hu(t) \tag{4.11}$$

Theorem 4.1 *The set $\mathscr{D}$ given by (4.8) is positively invariant with respect to the motion of the system (4.6) if and only if there exist matrices $H \in \mathbb{R}^{m\times m}$, $F_1 \in \mathbb{R}^{m\times n}$, and $F_2 \in \mathbb{R}^{m\times n}$ such that:*

$$KA_r = HK \tag{4.12}$$
$$\tilde{H}_c \ U \leq 0 \tag{4.13}$$

$$\tilde{H}_d \;\; U \leq \Delta \tag{4.14}$$

$$A_r \;\; \textit{is Hurwitz}. \tag{4.15}$$

Proof The result can be proved using a parallel approach to the one presented in [4, 21], but this time applied to the set (4.8) with respect to the system (4.11) while respecting the constraint on the rate by means of condition (4.14). It is also obvious that the set $\mathscr{D}$ given by (4.8) is positively invariant with respect to system (4.6) and the constraints on the control rate are respected if and only if the set Ω given by (4.2) is positively invariant with respect to the system (4.11) while the constraints on the control rate are respected. □

Remark 4.1 It is worth noting that the stability of the singular systems is realized by the stability condition (4.15). That is, the singular systems with the PD controller becomes admissible, while the constraints on both the control and its rate are respected for all initial conditions $x_o \in \mathscr{D}$.

Corollary 4.1 *For a given stabilizing gain matrix $F_1 \in \mathbb{R}^{m\times n}$ with full rank, there exists a solution $H \in \mathbb{R}^{m\times m}$, of equation $KA_r = HK$, if and only if:*

$$KA_r(\mathbb{I} - K^+K) = 0; \tag{4.16}$$

In this case, all the solutions are given by:

$$H = KA_rK^+ + Z(\mathbb{I} - KK^+), \tag{4.17}$$

where K^+ denotes any generalized inverse of K satisfying $KK^+K = K$, and $Z \in \mathbb{R}^{m\times m}$ is arbitrary.

Proof **Necessity**: Assume that Eqs. (4.16) and (4.17) hold true. Thence,

$$\begin{aligned} HK &= (KA_rK^+ + Z(\mathbb{I} - KK^+))K \\ &= KA_rK^+K + ZK - ZKK^+K \\ &= KA_rK^+K \\ &= KA_r \end{aligned}$$

then the obtained matrix H given by (4.17) is solution.

Sufficiency: Assume that there exist a matrix H solution of $KA_r = HK$, then:

$$\begin{aligned} KA_r - HK &= 0 \\ KA_r - HKK^+K &= 0 \\ KA_r - KA_rK^+K &= 0 \\ KA_r(\mathbb{I} - K^+K) &= 0 \end{aligned}$$

□

4.2.3 *Synthesis of the Constrained PD Controller*

In what follows, two methods to compute the PD controller are presented. The first one is based on an non-symmetrical nonlinear Riccati equation, which is very difficult to use. To overcome this difficulty, a second (heuristic) method is proposed, based on the solution of equation $XA + XBX = HX$, leading to an approximate solution.

4.2.3.1 Direct Procedure

This method consists in applying directly the results of the Corollary 4.1. The different steps of this technique are summarized by the following algorithm.

Algorithm 4.1

- Step 1: Compute matrix F_2 such that matrix $E + BF_2$ is invertible. Let $G_1 = (E + BF_2)^{-1}$.
- Step 2: Apply any pole assignment technique to compute a full rank matrix F_1 such that matrix A_r is Hurwitz while condition (4.16) is respected.
- Step 3: Compute matrix H in (4.17) using an arbitrary matrix Z.
- Step 4: If conditions (4.13)–(4.14) are satisfied, then a valid solution is already obtained; otherwise return to Step 3 (changing matrix Z) or to Step 1 (changing matrix F_2).

Note that the convergence of this algorithm is not guaranteed. Matrices Z and F_2 are "tuning" parameters and can be changed randomly. This explains why other synthesis methods are studied in the sequel of this chapter. Further, this direct procedure does not need that matrix H, computed with (4.17), be a diagonalizable matrix. Besides, if the augmentation technique is used, matrices B_a, F_{a1}, F_{a2}, K_a denote the corresponding augmented matrices to B, F_1, F_2, K (see the example).

Example: Direct procedure
Consider the singular system described by:

$$A = \begin{bmatrix} 1 & -1 & 0 \\ 0 & 1 & -1 \\ 1 & -2 & 0 \end{bmatrix}, \quad B = \begin{bmatrix} 0 & 0 \\ 1 & 0 \\ 0 & 1 \end{bmatrix}, \quad E = \begin{bmatrix} 1 & 0 & 0 \\ 0 & 1 & 1 \\ 0 & 0 & 0 \end{bmatrix},$$

which is subject to asymmetric constraints, corresponding to vectors:

$$U = [20 \;\; 8 \;\; 15 \;\; 8]^T, \quad \Delta = [130 \;\; 40 \;\; 130 \;\; 50]^T.$$

The finite generalized eigenvalues of (A, E) are -1 and 0.5: This implies that the open-loop system is not admissible. With matrix:

$$F_2 = \begin{bmatrix} -1 & -3 & -2 \\ 2 & -5 & -7 \end{bmatrix},$$

we chose to place the spectrum $\{-2, -3, -4\}$ and we obtain:

$$F_1 = \begin{bmatrix} -24 & 10 & -1 \\ 1 & -2 & 2 \end{bmatrix}, \; A_r = \begin{bmatrix} 1 & -1 & 0 \\ 17.88 & -8 & 1.77 \\ -12.77 & 6 & -1.55 \end{bmatrix}, \; K = \begin{bmatrix} 5.11 & -3 & 1.22 \\ -1 & 2 & 0 \end{bmatrix},$$

$$KA_r(\mathbb{I} - K^+K) = \begin{bmatrix} -2.97 & -1.48 & 8.78 \\ 1.68 & 0.84 & -4.96 \end{bmatrix} \neq \odot,$$

At this step, condition (4.16) is difficult to satisfy and can not be directly obtained. Henceforth, an alternative approach is proposed.

4.2.3.2 Non-symmetric Riccati Equation

Initially, it is possible to rewrite Eq. (4.12) by using relations (4.9)–(4.10), leading to the following equation to be solved:

$$MXLX + MXN_1 + N_2X + N_3 = 0 \tag{4.18}$$

where the variables in the equation are:

$$\begin{aligned} X &= F_1 \\ L &= G_1B \\ N_1 &= G_1A \\ N_2 &= -F_2G_1AG_1B - HM \\ N_3 &= HF_2G_1A - F_2G_1AG_1A \end{aligned} \tag{4.19}$$

Solution of this non-symmetric Riccati equation gives a direct method to solve the problem [1]. Unfortunately, as it is well known, this equation is rather difficult to solve exactly, especially when M is a singular matrix (as it will be shown latter, this is the case in the problem studied in this chapter). The work of [15] discussed this type of Riccati equations, giving some standard solutions.

4.2.3.3 Approximate Solution

Given the numerical difficulties of solving the Eq. (4.18) obtained with the direct approach, an approximate method is now proposed: The synthesis of the proposed controller based on the results of Theorem 4.1 is developed by using the so-called inverse procedure studied by [2]. Recall the following matrices:

$$
\begin{aligned}
G_1 &= (E + BF_2)^{-1} \\
M &= \mathbb{I}_m - F_2 G_1 B, \\
A_r &= G_1 (A + BF_1) \\
K &= MF_1 - F_2 G_1 A
\end{aligned} \tag{4.20}
$$

First, we are going to show that matrix M is singular, so that the pseudo inverse must be used: Assume that M^{-1} exists; by using the inverse lemma,

$$
(L + NCD)^{-1} = L^{-1} - L^{-1}N(C^{-1} + DL^{-1}N)^{-1}DL^{-1}, \tag{4.21}
$$

one can write: $M^{-1} = \mathbb{I}_m + F_2(G_1^{-1} - BF_2)^{-1}B = \mathbb{I}_m - F_2E^{-1}B$, so M is singular.

According to Lemma 1.5, Eq. (4.9) has a solution in the variable F_1 if and only if

$$
(\mathbb{I} - MM^+)(K + F_2G_1A) = 0. \tag{4.22}
$$

Further, one solution is given by:

$$
F_1 = M^+ (K + F_2 G_1 A), \tag{4.23}
$$

where M^+ denotes the Moore-Penrose inverse (pseudo inverse) of matrix M. Now, matrix A_r can be arranged as follows:

$$
A_r = G_1(\mathbb{I} + BM^+F_2G_1)A + G_1BM^+K, \tag{4.24}
$$

consequently, matrix A_r can be developed as

$$
\begin{aligned}
A_r &= \hat{A} + \hat{B}K \\
\hat{A} &= G_1(\mathbb{I} + BM^+F_2G_1)A \\
\hat{B} &= G_1BM^+,
\end{aligned} \tag{4.25}
$$

where $\hat{A} \in \mathbb{R}^{n\times n}$, $\hat{B} \in \mathbb{R}^{n\times m}$, and $M \in \mathbb{R}^{m\times m}$.

Applying Theorem 4.1 to the equation ($K\hat{A} + \hat{B}K = HK$) leads to a solution K as given above. Once the solution is obtained, F_1 can be computed according to Eq. (4.23).

It is worth noting that the use of the results of [2] requires that matrix $\hat{A}$ possesses at least $n - m$ stable eigenvalues. However, this condition is not restrictive since one can use the augmentation technique detailed in [4], as presented earlier. In order to summarize the necessary steps to apply for the synthesis of the controller, the following algorithm is given.

Algorithm 4.2

- Step 1: Compute matrix F_2 such that matrix $E + BF_2$ is invertible. Let $G_1 = (E + BF_2)^{-1}$
- Step 2: Compute matrices $\hat{A}$ and $\hat{B}$ given by (4.25), using the pseudo inverse of matrix M.
- Step 3: Choose a matrix $H \in \mathbb{R}^{m\times m}$ according to conditions (4.13)–(4.14) and conditions (1.37)–(1.40). Note that to determine matrix H, one can use the linear programming approach given in [21].
- Step 4: If matrix $\hat{A}$ does not admit $n - m$ stable eigenvalues, either return to Step 1, selecting a new matrix F_2, or augment B to make it square [4].
- Step 5: Compute the solution K of the algebraic equation (4.12).
- Step 6: Compute matrix F_1 from Eq. (4.23).
- Step 7: Compute the test condition $\varepsilon = (\mathbb{I} - M M^{+})(K + F_2 G_1 A)$: if ε is "small," then stop; otherwise repeat from Step 1, changing matrix F_2.

Remark 4.2 The proposed approach, which is frequently used in the literature [10], is based on selecting first F_2 and has the advantage of giving a "tuning" parameter for the designer. Of course, there are many matrices F_2 that fulfill the requirement that $E + BF_2$ is invertible, since $rank(E) = r$. In practical problems, the designer just selects one adequate matrix F_2 for the problem at hand.

4.2.4 System Augmentation Technique

In order to present an alternative way to design the required controllers of Theorem 4.1, one can use the following approach, based on relaxing the equalities of the algorithm:

First, it is possible to rewrite the system (4.1) under the equivalent form:

$$E\dot{x}(t) = Ax(t) + B_a sat(w(t)), \tag{4.26}$$

with matrix B_a corresponding to an augmentation of B to get a square matrix, given by:

$$B_a = \begin{bmatrix} B \ \odot \end{bmatrix},$$

where $\odot \in R^{n\times n-m}$ is a null matrix.

This augmentation is carried out through the introduction of a vector $v(t)$ made of $n - m$ fictitious entries, with their fictitious constraints given by: $-e_2 \leq v \leq e_1$ and $-\varphi_2 \leq \dot{v} \leq \varphi_1$. With this augmentation, the new control law becomes:

$$w(t) = \begin{bmatrix} u(t) \\ v(t) \end{bmatrix}, \tag{4.27}$$

with the feedback laws, $w(t) = K_a x(t)$ and $v(t) = K_{co}x(t)$. Define the real feedback law K_a, the new vector of constraints on the control magnitude U_a, and the new vector of constraints on the control magnitude Δ_a as follows:

$$K_a = \begin{bmatrix} K \\ K_{co} \end{bmatrix}; \quad U_a = \begin{bmatrix} U^a_{max} \\ U^a_{min} \end{bmatrix}; \quad \Delta_a = \begin{bmatrix} \Delta^a_{max} \\ \Delta^a_{min} \end{bmatrix},$$

where

$$U^a_{max} = \begin{bmatrix} u_{max} \\ e_1 \end{bmatrix}; \quad U^a_{min} = \begin{bmatrix} u_{min} \\ e_2 \end{bmatrix}; \quad \Delta^a_{max} = \begin{bmatrix} \Delta_{max} \\ \varphi_1 \end{bmatrix}, \quad \Delta_{min} = \begin{bmatrix} \Delta_{min} \\ \varphi_2 \end{bmatrix}. \tag{4.28}$$

Note that the system in closed loop given by the augmented control $w(t)$ remains the same as (4.6). On the other hand, the set of admissible constraints becomes with this augmentation:

$$\Omega_a = \{w \in \mathbb{R}^n / - U^a_{min} \leq w \leq U^a_{max}\}. \tag{4.29}$$

It is worth noting that this technique introduces new degrees of freedom with the variables e_i and φ_i, without modifying the system, which is simply augmented. Thus, the main advantage of the proposed technique is that it is possible to increase the size of the set of positive invariance by just selecting e_i and φ_i to be large.

The same development can be followed to obtain the closed-loop system (4.6), with augmented matrices F_1, F_2, M, H.

Suppose temporarily that the algebraic equations (4.12) was not exactly satisfied, and the error is restricted in an uncertain term D, as follows:

$$D = K_a A_r - H K_a. \tag{4.30}$$

By virtue of the new algebraic equation (4.30), the control dynamic is now given by

$$\begin{aligned} \dot{u}(t) &= (H K_a + D)x(t) \\ &= (H + L)u(t), \end{aligned} \tag{4.31}$$

where $L = DK_a^{-1}$ and the matrix K_a is assumed here to be non-singular. With this uncertain term L on the algebraic equation, an uncertain dynamic control is obtained. Assume now that the uncertain terms L are bounded by a known positive matrix $\Psi \in \mathbb{R}^{2n \times 2n}$:

$$\tilde{L}_d \leq \Psi. \tag{4.32}$$

In this case, according to [22], conditions (4.13)–(4.14) can be replaced with the following ones:

$$\begin{cases} \tilde{H}_c U_a + \Psi U_a \leq 0, \\ \tilde{H}_d U_a + \Psi U_a \leq \Delta_a. \end{cases} \tag{4.33}$$

Thus, the algorithm corresponding to this technique is the following:

Algorithm 4.3

- Step 1: Augment the system inputs with fictitious input vector $v(t)$ and the corresponding fictitious constraints e_1, e_2, φ_1, and φ_2.
- Step 2: Select matrix $F_2 \in \mathbb{R}^{n\times n}$ such that matrix $E + BF_2$ is invertible, let $G_1(E + BF_2)^{-1}$.
- Step 3: Compute matrices $\hat{A}$ and $\hat{B}$ given by (4.25) by using the pseudo inverse of matrix M.
- Step 4: Select matrices $H \in \mathbb{R}^{n\times n}$ and $\Psi \in \mathbb{R}^{2n\times 2n}$ according to the conditions (4.33) and (1.37)–(1.40).
- Step 5: Compute the solution K_a of the algebraic equation (4.12).
- Step 6: Compute matrix F_1, using Eq. (4.23).
- Step 7: In order to test convergence to a solution:

  (i) Compute the induced matrices A_r, K_a by using (4.20) and the obtained matrix F_1.
  (ii) Calculate the test conditions $\varepsilon_1 = norm((\mathbb{I} - MM^{+})(K_a + F_2G_1A))$ and $\varepsilon_2 = \tilde{L}_d - \Psi$.
  (iii) If ε_1 is "small" and $\varepsilon_2 < 0$, then stop; otherwise:
    - repeat from Step 2, selecting a new matrix F_2,
    - or repeat from Step 3 selecting new matrices H and/or Ψ.

Remark 4.3 • Note that to select matrix H in Step 3, it is possible to use the linear programming method given by [21].

- With the proposed method, the solution K_a of (4.12) obtained in Step 4 is always invertible, because $K_a = [K^T \quad K_{co}^T]^T$. In this case, $H = K^{-1}(A + BK)K$ and $A + BF = A + B_aK_a$.

Example 4.1 Approximate solution with augmentation.

Consider the following singular continuous-time system (studied by [7] and references therein):

$$A = \begin{bmatrix} 0 & 1 & 0 \\ -1 & 3 & 3 \\ -1 & 2 & 3 \end{bmatrix}; \; B = \begin{bmatrix} 0 & 0.2 \\ 1 & 0 \\ 0.9 & 0.8 \end{bmatrix}; \; E = \begin{bmatrix} 1 & 0 & 0 \\ 0 & 1 & 1 \\ -1 & 1 & 1 \end{bmatrix};$$

which is subject to asymmetric constraints, corresponding to vectors:

$$U = [5 \;\; 1 \;\; 1 \;\; 5]^T$$

$$\Delta = [5.5 \;\; 25.5 \;\; 21.5 \;\; 5.5]^T$$

It is easy to see that the finite generalized eigenvalues of (A, E) are 0.383 and 2.618: This implies that the open-loop system is not admissible. In order to relax the Algorithm 4.2, the technique of augmentation can be used in Step 4. For this, matrix B should become square of the form $B_a = [B \ \ \odot]$. This means that a fictitious input is added, together with the corresponding fictitious constraints on the control magnitude $(-e_1 \le u_3 \le e_2)$ and the control rate $(-\varphi_1 \le \dot{u}_3 \le \varphi_2)$. Thus, the constraint vectors become:

$$U_a = [5 \ \ 1 \ \ e_1 \ \ 1 \ \ 5 \ \ e_2]^T$$

and

$$\Delta_a = [5.5 \ \ 25.5 \ \ \varphi_1 \ \ 25 \ \ 5.5 \ \ \varphi_2]^T .$$

The Algorithm 4.2 was run selecting matrix F_2 randomly, until an acceptable error is obtained in Step 7. The following values of matrices M, $\hat{A}$ and $\hat{B}$ were obtained:

$$F_2 = \begin{bmatrix} 0.0048 & 0.0036 & -0.0018 \\ -0.0095 & 0.0009 & 0.0067 \\ 0.0055 & 0.0031 & 0.0006 \end{bmatrix}; \quad M = \begin{bmatrix} 0.9112 & 0.8484 & 0 \\ 0.0911 & 0.0848 & 0 \\ -0.0425 & 0.3915 & 1.0000 \end{bmatrix}$$

$$\hat{A} = \begin{bmatrix} 0.0002 & 0.9996 & -0.0005 \\ -1.0000 & 3.0000 & 3.0000 \\ -0.5000 & 2.0000 & 2.0000 \end{bmatrix}; \quad \hat{B} = 10^{15} \begin{bmatrix} -0.0000 & 0.0000 & -0.0000 \\ 0.2680 & -2.6802 & 0.0000 \\ -0.2680 & 2.6802 & -0.0000 \end{bmatrix}.$$

Note that $\sigma(\hat{A}) = \{4.8140;\ 0.0931 + 0.3086i;\ 0.0931 - 0.3086i\}$: That is, none of the eigenvalues of matrix $\hat{A}$ is stable. To use the technique based on the solution of the algebraic equation of [2], the following matrix H, that satisfies all the required conditions of positive invariance and fulfilling the constraints on the control magnitude and rate, was selected: (4.13)–(4.14):

$$H = \begin{bmatrix} -4 & 0.01 & 0.1 \\ 0 & -5 & 0.01 \\ 0 & 0 & -3 \end{bmatrix}$$

The fictitious constraints were also chosen to be large ($e_1 = 10$, $e_2 = 10$, $\varphi_1 = 30$ and $\varphi_2 = 30$). Matrix K follows together with matrix F_1 by using (1.42) and (4.23), respectively. Algorithm 4.2 was implemented using usual software, obtaining the following result:

$$K = 10^{-4}\begin{bmatrix} 0.0628 & 0.0628 & -0.2199 \\ 0.0063 & 0.0063 & -0.0220 \\ 0.5660 & -0.4340 & 0.2689 \end{bmatrix};$$

$$F_1 = \begin{bmatrix} -0.4949 & 1.7329 & 1.8100 \\ 0.5268 & -1.8513 & -1.9298 \\ -0.2305 & 0.8098 & 0.8423 \end{bmatrix};$$

Finally, to check the applicability of this synthesis method, matrices A_{re} and K_e were computed from matrix F_1 by using (4.20).

The obtained effective matrix in the closed-loop A_{re} is as follows:

$$A_{re} = \begin{bmatrix} 0.0002 & 0.9996 & -0.0005 \\ -92.5932 & 322.0706 & 337.1382 \\ 91.0932 & -317.0706 & -332.1382 \end{bmatrix}$$

Computing the spectrum of the effective matrix A_{re} gives

$$\sigma(A_{re}) = \{-8.6439; -0.7118 + 2.1068i; -0.7118 - 2.1068i\}.$$

It is possible to notice the difference between the pre-assigned spectrum and the obtained spectrum, due to the error of the algebraic equation. The condition of the use of the pseudo inverse (4.22) can also be checked: $\|(\mathbb{I} - M M^{+})(K + F_2 G_1 A)\| = 3.0512 \times 10^{-6}$, giving an acceptable error.

Finally, the norm of the error of the algebraic equation can also be computed, with the gain matrix K_e: $\|K_e\, A_{re} - H\, K_e\| = 3.58 \times 10^{-2}$. The effective gain matrices F_1 and K_e obtained with this approach are acceptable, even if the auxiliary matrices $\hat{B}$ and K are not satisfactory: A numerical dilemma appears between the value of the norm tests and the norm of matrix K.

The evolution of the state of system (4.6), the trajectory of the control, and the evolution of the control rate are presented, from the initial condition $x_o = [200\ 200\ \ -80]^T$ in Figs. 4.1, 4.2, and 4.3, respectively. Even if this approach is based on an approximate technique, the obtained set of positive invariance $\mathscr{D}$ is very large, which allows one to take a large initial value to plot these figures.

4.3 Extension to Non-singular Systems

This section shows that the obtained results can be easily extended to the control by proportional and derivative state feedback of non-singular linear systems with constraints on the control and its rate. Thus, the following linear system is considered in this subsection:

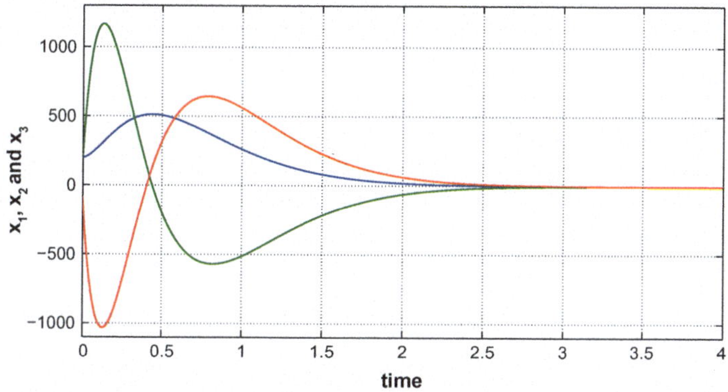

Fig. 4.1 States evolution

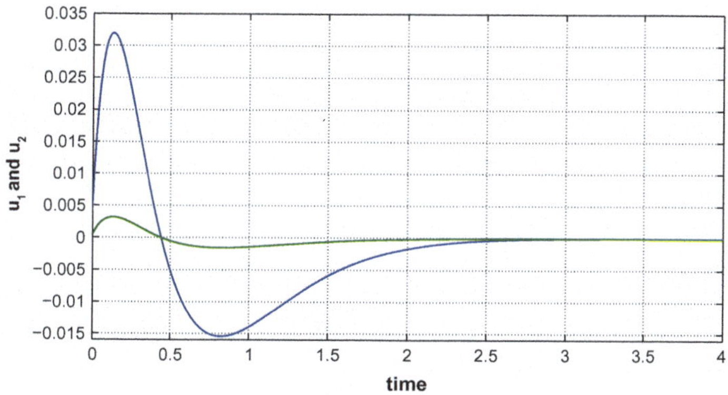

Fig. 4.2 Trajectory of the control

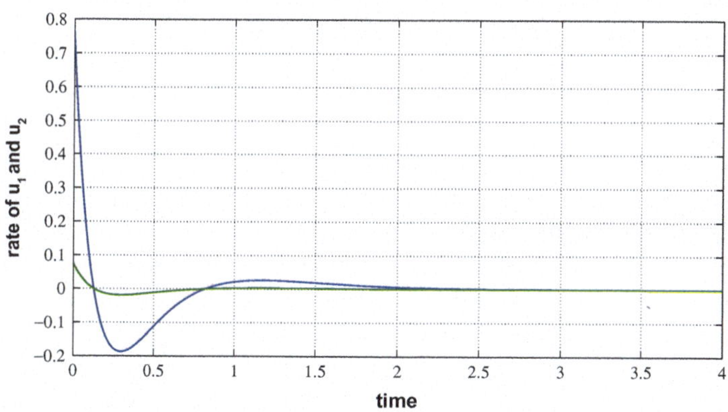

Fig. 4.3 Control rate evolution

$$\begin{aligned}\dot{x}(t) &= Ax(t) + Bu(t) \\ x(0) &= x_o\end{aligned} \tag{4.34}$$

where the control and its rate are constrained as in (4.2)–(4.3), and it is assumed that matrix A has $n - m$ stable eigenvalues and the pair (A, B) is stabilizable.

The objective in this case is to design a stabilizing state and state derivative feedback controller (PD):

$$u(t) = F_1x(t) - F_2\dot{x}(t), \tag{4.35}$$

where $F_1 \in \mathbb{R}^{m\times n}$; $F_2 \in \mathbb{R}^{m\times n}$, that respects the constraints on the control (4.2) and its rate (4.3).

To get the solution for non-singular systems, it is just necessary to replace E by the identity in the results for singular systems, obtaining:

$$\begin{aligned}G_1 &= (\mathbb{I} + BF_2)^{-1} \\ M &= \mathbb{I}_m - F_2G_1B\end{aligned} \tag{4.36}$$

In this case, it is not necessary to use the pseudo inverse of matrix M, as M^{-1} is well defined. Using the inversion Lemma, it is obtained that $M^{-1} = \mathbb{I}_m + F_2B$. Using the expression $(\mathbb{I} + BF_2)^{-1}B(\mathbb{I}_m + F_2B) = B$, the following developments are obtained:

$$\begin{aligned}\hat{A} &= ((\mathbb{I} + BF_2)^{-1} + (\mathbb{I} + BF_2)^{-1}B(\mathbb{I}_m + F_2B)F_2G_1)A = A, \\ \hat{B} &= (\mathbb{I} + BF_2)^{-1}B(\mathbb{I}_m + F_2B) = B,\end{aligned} \tag{4.37}$$

Once F_2 is selected, F_1 is given by:

$$F_1 = (\mathbb{I}_m + F_2B)(K + F_2A), \tag{4.38}$$

where matrix K is the solution of the algebraic equation

$$KA + KBK = HK, \tag{4.39}$$

that can be solved using the result of [2].

Thus, the Algorithms 4.2 and 4.3 can be followed to design the proposed controller, with the necessary modifications discussed above. Of course, in this case, the design method is exact since it is based directly on matrices A and B.

4.4 Conclusion

This chapter has presented necessary and sufficient conditions for positive invariance of a set $\mathscr{D}$ with respect to continuous-time singular linear system with constraints on the control and its rate, under state PD control. Practical algorithms for the synthesis of this controller have also been proposed by solving the algebraic equation $XA + XBX = HX$. An alternative augmentation-based algorithm was deduced, with the advantages of having additional degrees of freedom that can enlarge the set of positive invariance. An illustrative example has been presented, to show the applicability of the approach. The extension to non-singular systems has also been proposed.

References

1. H. Abou-Kandil, G. Freiling, V. Ionescu, G. Jank *Matrix Riccati equations in control and systems theory*. (Basel, Birkhauser, 2003)
2. A. Benzaouia, Resolution of equation $XA + XBX = HX$ and the pole assignment problem. IEEE Trans. Aut. Control **39**(10), (1994)
3. A. Benzaouia, C. Burgat, Regulator problem for linear discrete-time systems with non-symmetrical constrained control. Int. J. Control **48**, 2441–2451 (1988)
4. A. Benzaouia, A. Hmamed, Regulator problem for linear continuous systems with nonsymmetrical constrained control. IEEE Trans. Aut. Control **38**(10), 1556–1560 (1993)
5. A. Benzaouia, A. Hmamed, F. Tadeo and F. Mesquine, Regulation of linear singular systems under constrained control and rate, in *9th International Conference on Science and Techniques of Automation Control and Computer Engineering* (Tunisia)
6. A. Benzaouia, M. Darouach, A. Hmamed, Solution of equation $XA + XBXE = HXE$ and the pole assignment for singular systems. IMA J. Math. Control Inform. **29**(3), 343–356 (2012)
7. M. Chaabane, O. Bachelier, M. Souissi, D. Mehdi, Stability and stabilization of continuous descriptor systems an LMI approach. MPE **2006**, (2006) (Article ID 39367)
8. D. Chu, M. Malabre, Numerically reliable design for proportional and derivative state feedback decoupling controller. Automatica **38**, 2121–2125 (2002)
9. L. Dai, Observers for discrete singular systems. IEEE Trans. Aut. Control **33**(2), 187–191 (1988)
10. L. Dai, *Singular control systems* (Springer Verlag, Berlin, Lecture Notes in Control and Information Sciences, 1989)
11. M. Darouach, Solution to sylvester equation associated to linear descriptor. Syst. Control Lett. **55**(10), 835–838 (2006)
12. M. Darouach, M. Boutayeb, Design of observes for descriptor systems. IEEE Trans. Aut. Control **40**(7) 1323–1327 (1995)
13. M. Darouach, M. Zasadzinski, M. Hayar, Reduced-order observer design for descriptor systems with unknown inputs. IEEE Trans. Aut. Control **41**(7), 1068–1072 (1996)
14. M.B. Estrada, M. Malabre, Proportional and derivative state feedback decoupling of linear systems. IEEE Trans. Autom. Control **45**, 730–733 (2000)
15. G. Freiling, G. Jank, Non-symmetric Riccati equations. Zeitschrift fur Analysis und ihre Anwendungen **14**, 259–284 (1995)
16. C. Georgiou, N.J. Krikelis, A design approach for constrained regulation in discrete singular systems. Syst. Control Lett. **17**, 297–304 (1991)
17. J. Kaustky, N.K. Nichols, E.K.W. Chu, Robust pole assignment in singular systems. Linear Algebra Appl. **121**, 9–37 (1989)

18. V.X. Le, Synthesis of proportional-plus-derivative feedbacks for systems. IEEE Trans. Autom. Control **37**, 672–675 (1992)
19. F.L. Lewis, A survey of linear singular systems. Circuits Syst. Signal Process. **5**, 3–36 (1986)
20. C. Lin, Q.G. Wang, T.H. Lee, Robust normalization and stabilization of uncertain descriptor systems with norm bounded perturbations. IEEE Trans. Autom. Control **50**, 515–520 (2005)
21. F. Mesquine, F. Tadeo, A. Benzaouia, Regulator problem for linear systems with constraints on the control and its increments or rate. Automatica **40**(8), 1378–1395 (2004)
22. F. Mesquine, F. Tadeo, A. Benzaouia, Regulator constrained control and rate problem for linear systems with additive disturbances, in *Proceedings of ACC*, (Boston, 2004)
23. R. Mukundan, W. Dayawansa, Feedback control of singular systems: proportional and derivative feedback of the state. Int. J. Syst. Sci. **14**, 615–632 (1983)
24. P.C. Muller, M. Hou, On the observer design for descriptor systems. IEEE Trans. Aut. Control **38**(11), 1666–1671 (1993)
25. M. Naib, A. Benzaouia, F. Tadeo, l_1 control using linear programming for systems with asymmetric bounds. Int. J. Control **78**, 1459–1465 (2005)
26. P.N. Paraskevopoulos, F.N. Koumboulis, Observes for singular systems. IEEE Trans. Aut. Control **37**(8), 1211–1215 (1992)

Chapter 5
Observer-Based Constrained Control

5.1 Introduction

In this chapter, state feedback based on the observer with input saturation is presented. The estimate of the state will be used for the state feedback to stabilize the closed loop system under constrained control. Both full order and reduced-order observers will be of interest.

In another hand, the problem of constructing an observer for the singular system is also under interest bellow. In fact, it is an active area for research since two decades [5, 6, 8–10, 16], and the references therein.

The objective in this chapter is to construct such an observer which further respects the constraints on the control obtained only with the estimation of the state. The earlier work on regular systems [11] is extended in this work to singular systems.

This chapter is organized in two sections. First, full order and reduced-order observer are designed for constrained control systems. Hence, conditions are given such that the reconstructed state used for the control does not destroy the positive invariance property of the admissible set for control. Second, these results are extended to the case of singular systems.

5.2 Observer-Based Constrained Control

5.2.1 Problem Statement

Let us consider the continuous-time system given by

$$\begin{aligned} \dot{x}(t) &= Ax(t) + Bu(t), \\ y(t) &= Cx(t), \end{aligned} \tag{5.1}$$

A. Benzaouia et al., *Saturated Control of Linear Systems*, Studies in Systems, Decision and Control 124, https://doi.org/10.1007/978-3-319-65990-9_5

where x represents the state vector in $\mathbb{R}^n$, u is the control constrained to lie in the set Ω given by:

$$\Omega = \{u \in \mathbb{R}^m / -u_{min} \le u \le u_{max}, \quad u_{min}, u_{max} \in \mathbb{R}^m_+\} \tag{5.2}$$

with $(m \le n)$. $y(t)$ is the output of the system in $\mathbb{R}^{p_o}$. A, B, and C are matrices of appropriate sizes. We note the vector U as:

$$U^T = \left[u_{max}^T \; ; \; u_{min}^T\right]. \tag{5.3}$$

Further, we assume that:

$$(A, B) \text{ is stabilizable and } \quad (A, C) \text{ is observable.} \tag{5.4}$$

Furthermore, the feedback is made on the reconstructed state $\hat{x}(t)$ with:

$$u(t) = K\hat{x}(t), \tag{5.5}$$

where the estimate of the state $\hat{x}(t)$ is given by:

$$\dot{\hat{x}}(t) = A\hat{x}(t) + Bu(t) - K_o(y(t) - \hat{y}(t)), \tag{5.6}$$

where K_o is the observer matrix. Moreover, we define the set $\mathscr{D}(\mathbb{I}, \omega_1, \omega_2)$ by:

$$\mathscr{D}(\mathbb{I}, \omega_1, \omega_2) = \{e \in \mathbb{R}^n / -\omega_2 \le e \le \omega_1; \omega_1, \omega_2 \in \mathbb{R}^n_+ - \{0\}\} \tag{5.7}$$

which represents the evolution set for the state estimation error given by:

$$e = x - \hat{x}. \tag{5.8}$$

The constrained continuous-time regulator problem via observed state feedback (CCRPOF) is stated as follows: How to compute matrices K and K_o in order to guarantee the asymptotic stability of system (5.1) under the assumption (5.4) with the constraint $u \in \Omega$.

5.2.2 *Observer-Based Controller Design*

Consider that the feedback is made on the observed state as follows:

$$u(t) = sat(K.\hat{x}(t)), \; K \in \mathbb{R}^{m\times n}, \; rank\{K\} = m, \tag{5.9}$$

in such a way that:

$$Re(\lambda_i(A + B K)) < 0, \ i = 1, \ldots, n \tag{5.10}$$

The estimation error is subject to the dynamics given by

$$\begin{aligned} \dot{e}(t) = \dot{x}(t) - \dot{\hat{x}}(t) &= Ax(t) + Bu(t) - (A\hat{x}(t) + Bu(t) - K_o Ce(t)) \\ &= (A + K_o C)e(t). \end{aligned}$$

We obtain an autonomous system of the form:

$$\dot{e}(t) = Me(t), \tag{5.11}$$

where $M = A + K_o C$.

If follows from Theorem 1.8 that the set $\mathscr{D}(\mathbb{I}, \omega_1, \omega_2)$ is positively invariant with respect to the system (5.11) if only if:

$$\tilde{M}_c \omega \leq 0, \ with \ \ \omega = \begin{bmatrix} \omega_1 \\ \omega_2 \end{bmatrix}. \tag{5.12}$$

The conclusion from this development is that if K_o is computed such that the condition (5.12) is fulfilled and

$$Re(\lambda_i(A + K_o C)) < 0, \ i = 1, \ldots, n. \tag{5.13}$$

The observed state converges to the real ones and the error does not leave the domain $\mathscr{D}(\mathbb{I}, \omega_1, \omega_2)$ because it is positively invariant.
Let us now study the control dynamics:

$$\begin{aligned} \dot{u}(t) &= K\dot{\hat{x}}(t) \\ &= K(A\hat{x}(t) + Bu - K_o Ce(t)) \\ &= (KA + KBK)\hat{x}(t) - KK_o Ce(t) \end{aligned} \tag{5.14}$$

If there exits a matrix $H \in \mathbb{R}^{m \times m}$ solution to the algebraic equation:

$$KA + KBK = HK, \tag{5.15}$$

(5.14) becomes

$$\begin{aligned} \dot{u}(t) &= HK\hat{x}(t) - KK_o Ce(t), \\ &= Hu(t) + De(t), \end{aligned} \tag{5.16}$$

where $D = -KK_o C$.
Applying Theorem 1.8 to this system leads to the following corollary.

Corollary 5.1 *The domain* $int\Omega$*, given by (5.2), is positively invariant with respect to the motion of system (5.16) if and only if there exists a matrix* $H \in \mathbb{R}^{m\times m}$ *such that:*

$$KA + KBK = HK$$

$$\tilde{H}_c U + \tilde{D}_d \omega < 0,$$

where $\tilde{H}_c$*,* $\tilde{D}_d$*,* U *and* ω*, are as defined above.*

Proof The proof is obvious from the development above. □

Corollary 5.2 *The system* (5.6) *with* K *and* K_o *satisfying, respectively, the conditions* (5.10) *and* (5.13)*, is asymptotically stable for all initial conditions in the domain* (5.2)–(5.7) *if the two following conditions hold:*

- *There exists a matrix* H *solution to the algebraic equation* $K\ A\ +\ K\ B\ K = H\ K$*, such that:*

$$\tilde{H}_c U + \tilde{D}_d \omega < 0, \tag{5.17}$$

-

$$\tilde{M}_c \omega \leqslant 0, \tag{5.18}$$

where $M = A + BK_oC$ *and* $D = -KK_oC$*.*

Proof The first condition implies that the control u is always admissible in the domain $int\mathscr{D}(\mathbb{I}, u_{max}, u_{min})$. Furthermore, the second condition implies that the error does not leave the domain $\mathscr{D}(\mathbb{I}, \omega_1, \omega_2)$. Hence, the linear behavior is always guaranteed, and by bearing in mind that matrices K and K_o are computed such that conditions (5.10) and (5.13) hold, we can see that the asymptotic stability of the system is achieved. □

Algorithm 5.1 Without loss of generality, we assume that the matrix A has $n - m$ non-null stable eigenvalues. If this is not the case, one can easily augment the system with fictitious input as presented above. Further, let $\{\zeta_i\}_{m+1\leqslant i\leqslant n}$ be the $n - m$ eigenvectors of A associated with the $n - m$ invariant (stable) eigenvalues of A.

- Step 1. Choose $\Lambda_o = \{\zeta_i, i = 1, ..., n / Re(\zeta_i) < 0\}$, the observer dynamics.
- Step 2. Use any pole placement techniques to compute K_o such that:

$$\sigma(A + K_oC) = \Lambda_o,$$

- Step 3. Check that

$$\tilde{M}_c \omega \leqslant 0,$$

  is satisfied. If so continue, or else go to step 1.

- Step 4. Choose a matrix $H \in \mathbb{R}^{m\times m}$ such that

$$\tilde{H}_c U \leqslant 0.$$

Compute the eigenvectors of H, i.e., θ_i, $i = 1, ...m$, such as $H\theta_i = \lambda_i\theta_i$, with $\lambda_i \notin \sigma(A)$ and $Re(\lambda_i) < 0$, and $B\theta_i \neq 0$ and ζ_i the associated closed loop eigenvectors given by:

$$\zeta_i = (\lambda_i \mathbb{I} - A)^{-1} B\theta_i, \qquad i = 1, ...m. \tag{5.19}$$

- Step 5. Compute

$$K = [\theta_1 \cdots \theta_m\ 0\ \dots\ 0]\ [\zeta_1\ \zeta_2\ \cdots\ \zeta_n]^{-1} \tag{5.20}$$

- Step 6. If $\tilde{H}_c U + \tilde{D}_d\omega < 0$ go to step 7, or else go to step 4.
- Step 7. Use K and K_o for the CCRPOF.

Example 5.1 Let us consider the continuous-time system as in (5.1), where:

$$A = \begin{bmatrix} -3 & 2 \\ 2 & -1 \end{bmatrix},\ B = \begin{bmatrix} 1.51 & 0 \\ 0 & 1.51 \end{bmatrix},\ C = \begin{bmatrix} 1 & 0 \end{bmatrix}.$$

The constraints on the control law are as follows:

$$u_{max} = \begin{bmatrix} 0.65 \\ 1 \end{bmatrix},\ u_{min} = \begin{bmatrix} 1 \\ 0.5 \end{bmatrix},$$

the error that can be tolerated on the observed state is taken as follows:

$$\omega_1 = \begin{bmatrix} 0.1 \\ 0.09 \end{bmatrix}, \omega_2 = \begin{bmatrix} 0.08 \\ 0.1 \end{bmatrix}.$$

The observer gain K_o is computed using a pole placement technique such that the condition (5.13) is fulfilled. In addition, we check that the condition (5.12) holds, which leads to:

$$K_o = \begin{bmatrix} -0,5 \\ -2,5 \end{bmatrix},$$

with the spectrum: $\sigma(M) = \{-3; -1.5\}$ with $M = A + K_o C$ and

$$\tilde{M}_c\omega = [-0.17, -0.05, -0.08, -0.05]^T \leqslant 0.$$

The matrix H solution to (5.15) is given by:

$$H = \begin{bmatrix} -0.9411 & 0.151 \\ 0.0695 & -1.0339 \end{bmatrix}$$

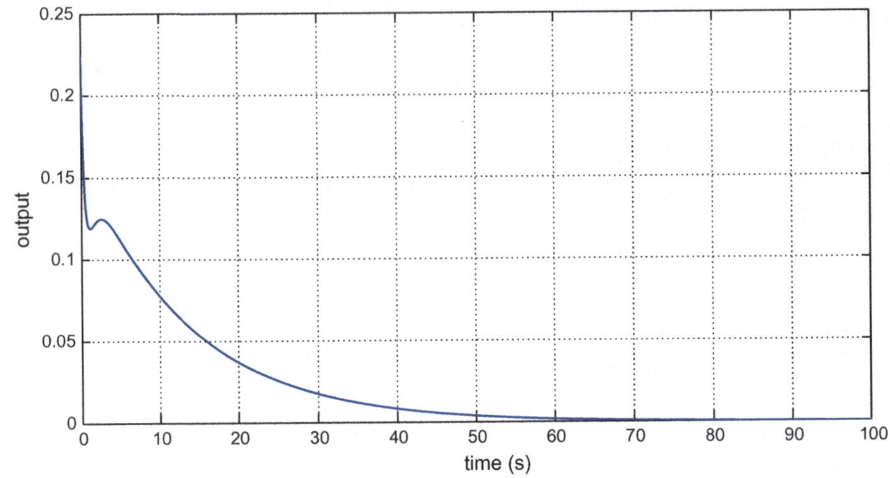

Fig. 5.1 The output of the closed loop system

with the feedback gain matrix:

$$K = \begin{bmatrix} 1.6522 & -1.6723 \\ -1.739 & 0.1089 \end{bmatrix},$$

obtained from the solution of (5.16) leading to the spectrum $\{-1.1, -0.875\}$ of the closed loop system. In order to check the validity of the condition (5.17), we compute the matrix:

$$D = -K\ K_o\ C = \begin{bmatrix} -3.354 & 0 \\ -0.597 & 0 \end{bmatrix},$$

and we obtain that:

$$\tilde{H}_c U + \tilde{D}_d \omega = [-0.192 - 0.94 - 0.53 - 0.387]^T \leqslant 0.$$

All the required conditions are satisfied, so one can conclude that the observed states converge to the real ones with no saturation on the control law and the closed loop is asymptotically stable (see the Figs. 5.1 and 5.2).

5.2.3 *Reduced-Order Observer-Based Constrained Control*

In the case where a part of the state is available, one has to reconstruct just the unobservable part of x. The presence of n_o linear combinations of the state at the output suggests that the remaining $n - n_o$ linear combinations may be reconstructed

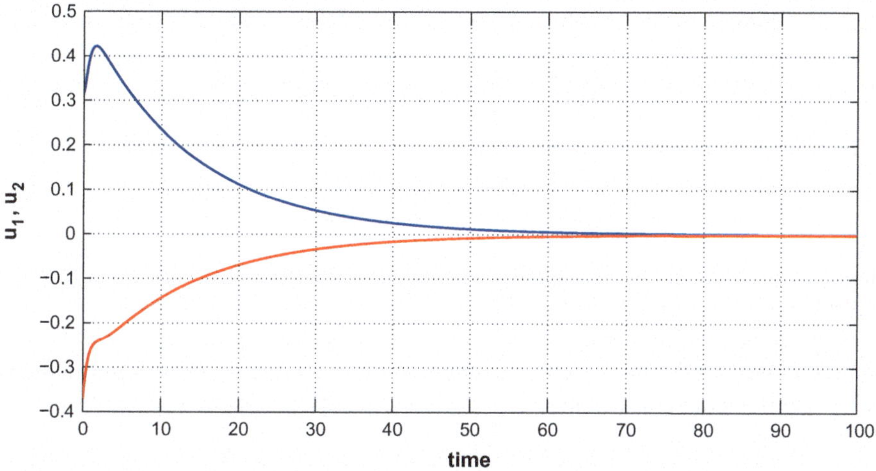

Fig. 5.2 Observer-based constrained control u_1(in *red*) u_2 (in *blue*) of the closed loop system

by an observer of order no greater that $n - n_o$ [15]. We generate $z(t) = Tx(t)$ as the $n - n_o$ linear state combination to be reconstructed. The state estimate in this case is given by:

$$\hat{x}(\cdot) = \begin{bmatrix} C \\ T \end{bmatrix}^{-1} \begin{bmatrix} y(\cdot) \\ z(\cdot) \end{bmatrix} = \begin{bmatrix} V & P \end{bmatrix} \begin{bmatrix} y(\cdot) \\ z(\cdot) \end{bmatrix} \tag{5.21}$$

where the matrix T is chosen such that the required inverse matrix exists. Since $Tz(\cdot)$ can not be measured exactly, it will be reconstructed from an auxiliary dynamical system of order $n - n_o$.

$$\dot{z}(t) = Dz(t) + Ny(t) + Gu(t) \tag{5.22}$$

In this section, we will present a controller-based reduced-order observer that takes into account the constraints on the control using positive invariance results [12]. Let us consider the linear constrained system:

$$\begin{cases} \dot{x}(t) = A\,x(t) + \;B\,u(t) \\ y(t) = C\;x(t) \end{cases} \tag{5.23}$$

where $x(t) \in \mathbb{R}^n$ is the state vector, $u(t) \in \mathbb{R}^m$ is the control vector, $y(t) \in \mathbb{R}^{p_o}$ is the output vector, A and B are constant matrices of appropriate dimension and (A, B) is controllable. It is assumed that A possesses at least $(n - m)$ stable eigenvalues. The control u is constrained in the set Ω defined by (5.2). Generally using a state feedback control:

$$u(t) = sat(K\,x(t)), \quad K \in \mathbb{R}^{m \times n}, \tag{5.24}$$

leads to a domain of linear behavior for the closed loop system that is given by:

$$\mathscr{D}(K, u_{min}, u_{max}) = \{x \in \mathbb{R}^n \backslash -u_{min} \leq K x \leq u_{max}\}, \tag{5.25}$$

and the closed loop system in this case:

$$\dot{x}(t) = (A + BK)x(t). \tag{5.26}$$

Hence, if the domain (5.25) is positively invariant, one guarantees the respect of the control constraints for all $t \geq 0$.

At this stage, our problem may be stated as finding matrices K, D, N, and G such that the closed loop system with the control $u(t) = sat(K\hat{x}(t))$ is asymptotically stable and the input constraints are respected. The observed state is given by:

$$\hat{x}(t) = \begin{bmatrix} C \\ T \end{bmatrix}^{-1} \begin{bmatrix} y(t) \\ z(t) \end{bmatrix} = \begin{bmatrix} V & P \end{bmatrix} \begin{bmatrix} y(t) \\ z(t) \end{bmatrix}, \tag{5.27}$$

where the matrices $V, \quad C, \quad T$, and P satisfy

$$V\ C\ +\ P\ T = \mathbb{I}. \tag{5.28}$$

Recall that the minimal order observer matrices, as proposed, are given by [15]

$$D = T\ A\ P, \quad N = T\ A\ V, \quad \text{and } G = T\ B, \tag{5.29}$$

which is equivalent to:

$$T\ A - N\ C = D\ T. \tag{5.30}$$

Matrix P is chosen to ensure asymptotic stability of the matrix D. In fact, matrix D defines the dynamics of the errors and this guarantees vanishing errors [11]. Indeed,

$$\begin{aligned}
\dot{\varepsilon}(t) &= \dot{z}(t) - T\dot{x}(t) \\
&= Dz(t) + Ny(t) + Gu(t) - T(Ax(t) + Bu(t)) \\
&= Dz(t) + NCx(t) - TAx(t) \\
&= Dz(t) - DTx(t) \\
&= D\varepsilon(t).
\end{aligned}$$

For the observation error, we define the field $\mathscr{D}(\mathbb{I}, \varepsilon_{max}, \varepsilon_{min})$ in which we allow the error $\varepsilon(t)$ to evolve:

$$\mathscr{D}(\mathbb{I}, \varepsilon_{max}, \varepsilon_{min}) = \left\{\varepsilon(t) \in \mathbb{R}^{n-n_o} / \varepsilon_{min} \leq \varepsilon \leq \varepsilon_{max}\right\}. \tag{5.31}$$

Further, define the reconstruction error as $e(t) = \hat{x}(t) - x(t)$. Note that, it is related to the observation error as follows:

$$
\begin{aligned}
e(t) &= V\ y(t) + P\ z(t) - x(t) \\
&= V\ Cx(t) + Pz(t) - (V\ C + P\ T)x(t) \\
&= P(z(t) - T\ x(t)) \\
&= P\ \varepsilon(t)
\end{aligned}
$$

Furthermore, one may prove that the control dynamics are as follows [12]:

$$
\begin{aligned}
\dot{u}(t) &= K\ \dot{\hat{x}}(t) \\
&= KP\dot{z}(t) + KVC\dot{x}(t) \\
&= KP(Dz(t) + Ny(t) + Gu(t)) + KVC(Ax(t) + Bu(t)) \\
&= KP(TAPz(t) + TAVy(t)) + (KPTB + KVCB)u(t) + KVCAx(t) \\
&= KPTA(Pz(t) + Vy(t)) + K(PT + VC)Bu(t) + KVCAx(t) \\
&= KPTA\hat{x}(t) + KBu(t) + KVCA(\hat{x}(t) - e(t)) \\
&= K(PT + VC)A\hat{x}(t) + KBK\hat{x}(t) - KVCAe(t) \\
&= (KA + KBK)\hat{x}(t) - KVCAe(t) \\
&= HK\hat{x}(t) - KVCAP\varepsilon(t) \\
&= Hu(t) + L_r\varepsilon(t)
\end{aligned}
$$

Therefore the system formed by the control $u(t)$ and the error $\varepsilon(t)$ is obtained as:

$$
\begin{bmatrix} \dot{u}(t) \\ \dot{\varepsilon}(t) \end{bmatrix} = \begin{bmatrix} H & L_r \\ 0 & D \end{bmatrix} \begin{bmatrix} u(t) \\ \varepsilon(t) \end{bmatrix} \tag{5.32}
$$

This background enables one to recall the theorem giving conditions for computing the controller that respects all the needed requirements:

Theorem 5.1 [12] *The field $\mathscr{D}(\mathbb{I}, u_{max}, u_{min}) \times \mathscr{D}(\mathbb{I}, \varepsilon_{max}, \varepsilon_{min})$ is positively invariant with respect to the trajectory of system (5.32) if and only if, there exists a matrix $H \in R^{m\times m}$ such that:*

$$
\begin{cases} H\ K &= K\ A + K\ B\ K \\ \widetilde{W}_c\ q_\varepsilon &\leq 0 \end{cases} \tag{5.33}
$$

where

$$
W = \begin{bmatrix} H & L_r \\ 0 & D \end{bmatrix};\ q_\varepsilon = \begin{bmatrix} u_{max} \\ \varepsilon_{max} \\ u_{min} \\ \varepsilon_{min} \end{bmatrix};\ L_r = -K\ V\ C\ A\ P \tag{5.34}
$$

for every pair $(u(0), \varepsilon(0)) \in \mathscr{D}(\mathbb{I}, u_{max}, u_{min}) \times \mathscr{D}(\mathbb{I}, \varepsilon_{max}, \varepsilon_{min})$.

To compute the feedback gain, the inverse procedure is used [2, 3]. Hence, matrix H satisfying all required conditions such that a solution exists is chosen and the feedback K is obtained as a solution to the equation:

$$
K\ A + K\ B\ K\ =\ H\ K. \tag{5.35}
$$

The controllers proposed here are shown to be robust with respect to parametric uncertainties within given sets for the system matrices. For more details about robustness and sensitivity of such controllers, the reader is refereed to [14].

Remark 5.1 Note here that all computation effort is handled off line. Choice of an adequate matrix H with all required conditions is studied in [13]. Solution of Eq. (5.35) is the detailed subject of the work [2].

5.2.4 Reduced-Order Observer Framework

Without loss of generality, we consider the class of systems for which matrix C is written such that

$$C = [\ \mathbb{I} \quad 0\ ] \tag{5.36}$$

In fact, all systems with matrix C of full row rank can be partitioned as:

$$C = C_r\ \bar{C}\ C_l, \text{ where } C_r \in \mathbb{R}^{p_o \times p_o},\ C_l \in \mathbb{R}^{n \times n} \tag{5.37}$$

where C_r and C_l are non-singular matrices and $\bar{C}$ is of the form (5.36). Hence, one can note that with an appropriate change of coordinates we return to the case where C is of the form (5.36). Partition now the system and the matrices P and V like [15]:

$$\dot{x}(t) = \begin{bmatrix} A_{11} & A_{12} \\ A_{21} & A_{22} \end{bmatrix} x(t) + \begin{bmatrix} B_1 \\ B_2 \end{bmatrix} u(t),\ \ A_{11} \in \mathbb{R}^{p_o \times p_o},\ \ A_{12} \in \mathbb{R}^{p_o \times (n-p_o)}.$$

For matrix observer, we obtain the following decomposition:

$$V = \begin{bmatrix} V_1 \\ V_2 \end{bmatrix},\ \ V_1 = \mathbb{I}_{p_o}$$
$$P = \begin{bmatrix} P_1 \\ P_2 \end{bmatrix},\ \ P_1 = 0_{p_o \times (n-p_o)}$$

In this case, we have:

$$[V\ \ P] = \begin{bmatrix} \mathbb{I}_{p_o} & 0 \\ V_2 & P_2 \end{bmatrix}$$

The parametrization of the observer is completely achieved by choosing any non-singular matrix P_2 and by computing V_2 such that $(A_{22} - V_2 A_{12})$ is stable. Hence, one can take $P_2 = \mathbb{I}_{n-n_o}$. Further, we decompose the feedback K as:

$$K = [K_1 \quad K_2] \tag{5.38}$$

Matrices L_r and D reduce to the following:

$$D = A_{22} - V_2 A_{12} \quad \text{and} \quad L_r = (K_1 + K_2 V_2) A_{12} \tag{5.39}$$

and the augmented system to

$$\begin{bmatrix} \dot{u}(t) \\ \dot{\varepsilon}(t) \end{bmatrix} = \begin{bmatrix} H & (K_1 + K_2 V_2) A_{12} \\ 0 & A_{22} - V_2 A_{12} \end{bmatrix} \begin{bmatrix} u(t) \\ \varepsilon(t) \end{bmatrix} \tag{5.40}$$

$$= W \begin{bmatrix} u(t) \\ \varepsilon(t) \end{bmatrix}. \tag{5.41}$$

It is worth noting here that matrix V_2 suffices to define the reduced-order observer.

Remark 5.2 1. The admissible initial conditions set for x_o and z_o, respectively, for the state space and the reduced-order observer are defined as:

$$-w_{2r} \leq z_o - T x_o \leq w_{1r} \tag{5.42}$$

$$-u_{min} \leq K P z_o + K V C x_o \leq u_{max} \tag{5.43}$$

this set is non-void since it contains always an admissible state given by

$$z_o = T x_o, \quad \text{such that} \quad K P T x_o + K V C x_o = K x_o$$

for an x_o satisfying

$$x_o \in \mathscr{D}(K, u_{max}, u_{min})$$

2. For systems with unconstrained control, the separation principle is used to design both the controller and the observer separately. But for constrained control systems, this is no longer valid. In this sense, conditions (5.17)–(5.18) in the full order case and (5.33) in the reduced-order case can be seen as a way to choose the observer such that the property of positive invariance of domain $\mathscr{D}(K, u_{max}, u_{min})$ is not destroyed.

Example 5.2 Consider the continuous-time system modeling the plane AFTI/F16 flying at the altitude of 3000 ft [17]:

$$A = \begin{bmatrix} 0 & 1 & 0 \\ 0 & -0.87 & 43.22 \\ 0 & 0.99 & -1.34 \end{bmatrix}, \quad B = \begin{bmatrix} 0 & 0 \\ -17.25 & -1.58 \\ -0.17 & -0.25 \end{bmatrix}, \quad C = \begin{bmatrix} 1 & 0 & 0 \\ 0 & 1 & 0 \end{bmatrix},$$

we assume that the constraints are given by:

$$u_{max} = \begin{bmatrix} 90 \\ 90 \end{bmatrix}, \quad u_{min} = \begin{bmatrix} 45 \\ 45 \end{bmatrix},$$

the set of accepted errors is given by (5.31) where:

$$\varepsilon_{max} = 2\varepsilon_{min} = 2.$$

One may choose to assign the spectrum $\{-0.5 \quad -6\}$ by matrix H as follows:

$$H = \begin{bmatrix} -0.5 & -0.1 \\ 0 & -6 \end{bmatrix}.$$

The solution of the algebraic equation (5.15) gives:

$$K = \begin{bmatrix} 5.42 & 0.84 & 0.87 \\ -36.38 & -1.85 & 18.46 \end{bmatrix}.$$

The reduced-order observer is fully determined by matrix V_2 as: $V_2 = [\ 0 \ \ 0.2004\ ]$, which satisfy the required conditions. Matrix D, in this case, reduces to a scalar -10, which is a stable eigenvalue defining the dynamic of the observer. The required condition to be checked is given by the vectors:

$$\begin{aligned} T &= \begin{bmatrix} -V_2 & \mathbb{I}_{n-p_o} \end{bmatrix} = [0 \quad -0.2004 \quad 1] \\ N &= T \;\; A \;\; V = [0 \quad -0.8394] \\ G &= TB = [3.2864 \quad 0.066]\,. \end{aligned}$$

Figures 5.3, 5.4, and 5.5 present respectively observer state, output, and control for $z_0 = 4$.

One can conclude that the close-loop system is asymptotically stable while all constraints are satisfied and the observed states converge to the real ones.

5.3 Constrained Observer-Based Control for Singular Linear Systems

5.3.1 Problem Formulation

In this section, we give the problem formulation related to singular linear system with constrained control studied in this chapter.

Consider the following system described by:

$$E\dot{x}(t) = Ax(t) + Bu(t) \tag{5.44}$$
$$y(t) = Cx(t) \tag{5.45}$$
$$x(0) = x_o$$

where $x \in \mathbb{R}^n$ is the state, $u \in \mathbb{R}^m$ is the control and $y \in \mathbb{R}^{p_o}$ the output with, $Rank(E) = r \leq n$ and $Rank(C) = n_o \leq n$. Assume that,

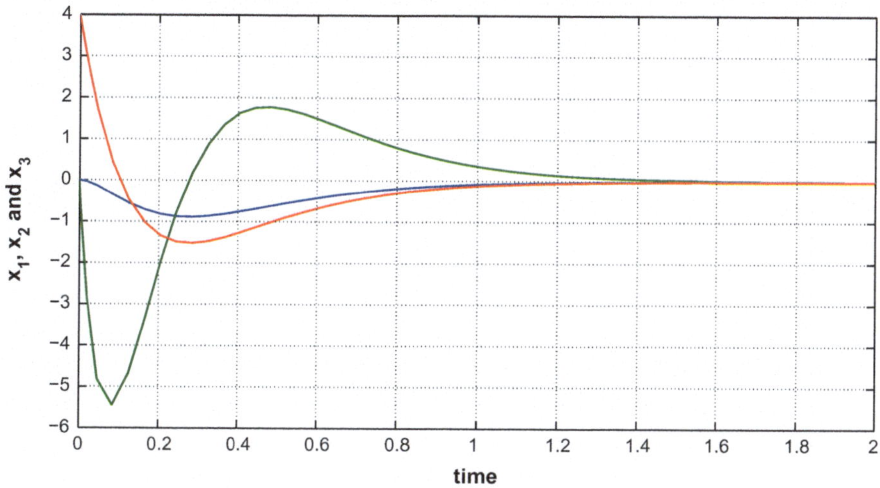

Fig. 5.3 Observer state evolution

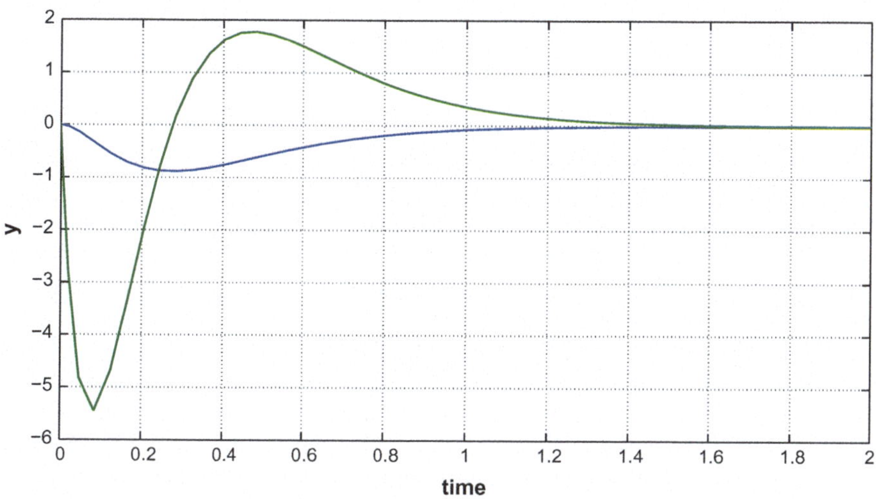

Fig. 5.4 System output evolution

(AS1) (E, A, B) is stabilizable,
(AS2) (E, A, C) is detectable and $m \leq r$.

The control is assumed here to be constrained in Ω given by (5.2). Generally, when the state is available, the control is given by:

$$u(t) = Fx(t), \ F \in \mathbb{R}^{m \times n}, \ rank(F) = m. \tag{5.46}$$

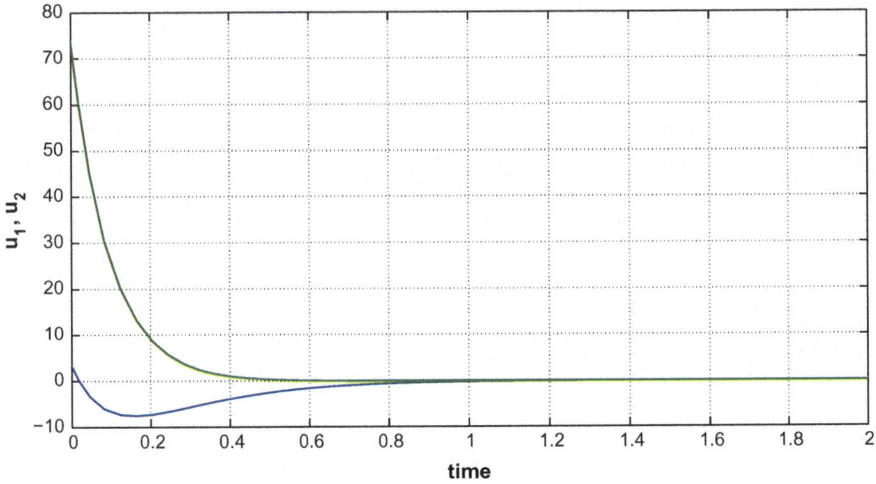

Fig. 5.5 Control evolution

The unsaturated system in closed loop is then obtained by:

$$\begin{aligned} E\dot{x}(t) &= (A + BF)x(t) \\ &= A_{cl}x(t). \end{aligned} \tag{5.47}$$

Assume that $rank(F) = m$ and F is stabilizing the system in closed loop (5.47), which means that the system in closed loop is admissible (regular and impulse free). With this system in closed loop, the induced constraint set on the state is given by,

$$\mathscr{D} = \left\{x \in \mathbb{R}^n | -u_{min} \leq Fx \leq u_{max}\right\}. \tag{5.48}$$

The main idea of the approach of positive invariance is to impose to the system trajectories to evolve only inside the region of linear behavior defined by (5.48) to have Eq. (5.47) valid. This representation of the singular system is used directly to design the stabilizing controller in the presence of constraints by using observers.

5.3.2 *Observer-Based Constrained Control for Singular Systems*

We assume in this subsection that the state is not available.
Assume that:
(AS3)

$$rank\begin{bmatrix} E \\ C \end{bmatrix} = n.$$

Consider the following reduced-order observer for the singular linear system:

$$\begin{aligned}\dot{z}(t) &= Dz(t) + Ny(t) + Gu(t) \\ \hat{x}(t) &= Pz(t) + Vy(t)\end{aligned} \tag{5.49}$$

where $z \in \mathbb{R}^{n-n_o}$ is the observer state, matrices D, N, G, P, V are constant matrices of appropriate size to be determined. Note the errors of observation and reconstitution as follows, respectively:

$$\varepsilon(t) = z(t) - TEx(t), \tag{5.50}$$

$$e(t) = \hat{x}(t) - x(t), \tag{5.51}$$

where matrix $T \in \mathbb{R}^{(n-n_o)\times n}$ is also to be computed. Recall that the dynamic error is given by: [7]

$$\dot{\varepsilon}(t) = D\varepsilon(t). \tag{5.52}$$

Since the objective of the observer is to converge to the state of the system, that is the errors tend to zero as the time goes to infinity, one can then assume that the error has to evolve inside pre-given limitations defined by:

$$-\varepsilon_{min} \leq \varepsilon \leq \varepsilon_{max}, \bar{\varepsilon} = \begin{bmatrix} \varepsilon_{max} \\ \varepsilon_{min} \end{bmatrix}. \tag{5.53}$$

Consider the following control,

$$u(t) = F\hat{x}(t). \tag{5.54}$$

The set of induced constraints on the estimation of the state is given by:

$$\mathscr{D}_e = \left\{\hat{x} \in \mathbb{R}^n | -u_{min} \leq F\hat{x} \leq u_{max}\right\}. \tag{5.55}$$

The unsaturated system in closed loop obtained with the observer is:

$$E\dot{x}(t) = Ax(t) + BF\hat{x}(t). \tag{5.56}$$

The following results presents the necessary and sufficient condition of existence of a reduced-order observer for the singular linear system (5.44), which respects the constraints on the control given by (5.55).

Theorem 5.2 *Assume that assumptions $AS1 - AS3$ are satisfied, then there exists a constrained observer-based controller for the singular system (5.44) realizing the* *positive invariance of the set $\mathscr{D}_e$ given by (5.55) with respect to the system (5.56) if and only if there exist matrices H, T, N, P, V, Γ, and a Hurwitz matrix D satisfying:*

$$\Gamma A + \Gamma B \Gamma E = H \Gamma E, \tag{5.57}$$

$$TA - DTE = NC, \tag{5.58}$$

$$PTE + VC = \mathbb{I}, \tag{5.59}$$

$$G = TB,$$

such that:

$$\tilde{H}_c U + \tilde{\Psi}_d \bar{\varepsilon} \leq 0, \tag{5.60}$$

$$\tilde{D}_c \bar{\varepsilon} \leq 0, \tag{5.61}$$

with $\Psi = \Gamma(EPD - AP)$ and $F = \Gamma E$.

Proof Compute the control dynamic by using Eqs. (5.58)–(5.59), the observer (5.49), the errors expressions (5.50)–(5.51) and $F = \Gamma E$:

$$\begin{aligned}
\dot{u}(t) &= F\dot{\hat{x}}(t) = FP\dot{z}(t) + FVC\dot{x}(t) \\
\dot{u}(t) &= FPDz(t) + FP(TA - DTE)x(t) + FPGu(t) + FVC\dot{x}(t) \\
&= FPDz(t) + FPT(Ax(t) + Bu(t)) - FPDTEx(t) + FVC\dot{x}(t) \\
&= FPDz(t) + F(PTE + VC)\dot{x}(t) - FPDTEx(t).
\end{aligned}$$

Recall that $PTE + VC = \mathbb{I}$, it follows:

$$\dot{u}(t) = FPDz(t) + F\dot{x}(t) - FPDTEx(t).$$

By using $F = \Gamma E$ and Eq. (5.56), one obtains:

$$\dot{u}(t) = \Gamma EPDz(t) + \Gamma(A - EPDTE)x(t) + \Gamma B \Gamma E\hat{x}(t).$$

By substituting by (5.50), it follows:

$$\dot{u}(t) = \Gamma EPD\varepsilon(t) + \Gamma(EPDTE + A - EPDTE)x(t) + \Gamma B \Gamma E\hat{x}(t).$$

Finally, the use of (5.51) leads to:

$$\dot{u}(t) = \Gamma(A + B\Gamma E)\hat{x}(t) - \Gamma Ae(t) + \Gamma EPD\varepsilon(t). \tag{5.62}$$

It is well known [7] that under Eqs. (5.50)–(5.51), there exists a link between the observation error and the reconstruction error, $e(t) = P\varepsilon(t)$. Taking account of the algebraic equation (5.57), it follows:

$$\dot{u}(t) = Hu(t) + \Psi\varepsilon(t). \tag{5.63}$$

Using an augmented state with Eqs. (5.52) and (5.63), the following system is obtained:

$$\dot{\zeta}(t) = \begin{bmatrix} H & \Psi \\ O & D \end{bmatrix} \zeta(t);\ \zeta = \begin{bmatrix} u \\ \varepsilon \end{bmatrix}. \tag{5.64}$$

Once the dynamic control is obtained, one can use again the result of [1] to system (5.64) to obtain the necessary and sufficient conditions guaranteeing that the control obtained with the observer always respects the constraints. These conditions are given by (5.60)–(5.61). It is worth noting that a necessary condition to realize this is

$$\tilde{H}_c\ U \leq 0 \text{ since } \tilde{\Psi}_d\ \bar{\varepsilon} \geq 0.$$

Further, since matrix D is stable, the error $\varepsilon(t)$ will tend to zero as the time goes to infinity while respecting its proper constraints since condition (5.61) is also satisfied. The choice of F such that the singular system in closed loop is regular and impulse free will ensure the asymptotic stability of the singular system with the observer. Note that asymptotically, system (5.56) will become exactly the same as (5.47). For this, generally, it is recommended that the dynamic of the error is faster than the dynamic of the singular systems in closed loop. □

Remark 5.3 In [7], a detailed solution of Eqs. (5.58)–(5.59) is presented. It is shown that matrices D and N can be selected as $D = TAP$ and $N = TAV$ to fulfill Eqs. (5.58)–(5.59). Using these expressions, matrix Ψ becomes, $\Psi = \Gamma(PT - \mathbb{I})AP$. For the particular case of linear systems, $E = \mathbb{I}$, $\Gamma = F$ and $\Psi = -FVCAP$, which was found by [11] for standard linear systems.

Example 5.3 In order to apply the results of Theorem 5.2, consider the following continuous-time singular linear system:

$$A = \begin{bmatrix} 1 & -1 & 0 \\ 0 & 1 & -1 \\ 1 & -2 & 0 \end{bmatrix},\quad B = \begin{bmatrix} 0 & 0 \\ 1 & 0 \\ 0 & 1 \end{bmatrix},\quad C = [0\ \ 0\ \ 1],$$

$$E = \begin{bmatrix} 1 & 0 & 0 \\ 0 & 1 & 1 \\ 0 & 0 & 0 \end{bmatrix},\ \text{with } S_\infty = \begin{bmatrix} 0 \\ -0.7071 \\ 0.7071 \end{bmatrix},$$

$$\Phi_0 = \begin{bmatrix} 0 & 0 \\ 1 & 0 \\ 0 & 1 \end{bmatrix}\quad \Phi_1 = \begin{bmatrix} 1 \\ 0 \\ 0 \end{bmatrix}.$$

The pencil $[A, E]$ has two finite generalized eigenvalues: $\sigma([A, E]) = \{0.5; -1\}$ and matrix E is singular with $rank(E) = 2$.

The control is assumed here to be constrained as (5.2), with:

$$U = \begin{bmatrix} u_{max} \\ u_{min} \end{bmatrix} = [30 \;\; 25 \;\; 40 \;\; 30]^T .$$

Choose matrix H according to assumptions (1.104)–(1.107) as follows:

$$H = \begin{bmatrix} -4 & 1 \\ 0 & -5 \end{bmatrix}.$$

The use of the solution method given in [9] leads to the following observer matrices:

$$D = \begin{bmatrix} -0.6 & 0.2 \\ -2 & -5 \end{bmatrix}, \quad P = \begin{bmatrix} 0.04 & -0.28 \\ 0 & -0.2 \\ 0 & 0 \end{bmatrix},$$

$$V = \begin{bmatrix} 0 \\ -1 \\ 1 \end{bmatrix}, \quad N = \begin{bmatrix} 15 \\ -90 \end{bmatrix}.$$

The a priori imposed constraints on the error are given by, $\bar{\varepsilon} = [3 \; 5 \; 4 \; 3]^T$. Matrix Ψ is computed as:

$$\Psi = \begin{bmatrix} -13.8537 & -41.5611 \\ 15.9960 & 47.9880 \end{bmatrix}.$$

Conditions (5.60)–(5.61) are then checked to be satisfied:

$$\tilde{H}_c g + \tilde{\Psi}_d \bar{\varepsilon} = \begin{bmatrix} -2.141 \\ -22.832 \\ -1.426 \\ -76.212 \end{bmatrix},$$

$$\tilde{\Psi}_c \bar{\varepsilon} = \begin{bmatrix} -0.8 \\ -17 \\ -1.8 \\ -9 \end{bmatrix}.$$

The pencil $[A + BF, E]$ has two finite generalized eigenvalues: $\sigma([A + BF, E]) = \{-4; -5\}$ while $\sigma(D) = \{-0.6929; -4.9071\}$.

5.4 Conclusion

Our interest, in this chapter, has been focused on the problem of observer-based controller design with constrained control. Condition of positive invariance applied to non-autonomous linear systems is then used in the design of an observer-based controller. An algorithm for such a design is given and applied in an illustrative example.

In addition, new necessary and sufficient conditions for the existence of a constrained observer-based controller for continuous-time singular linear system are presented. These conditions deal directly with the system matrices without any use of a transformation. Its design is also presented by using the solution of the algebraic equation $XA + XBXE = HXE$ given in [4]. The obtained controller with this approach guarantees that the system in closed loop is regular, impulse free and always has a linear comportment inside a region of positive invariance. The consequence is that the asymmetric constraints on the control, even by using an observer, are respected. An illustrative example is studied to show the applicability of this approach.

References

1. A. Benzaouia, A. Hmamed, Regulator problem for linear continuous-time systems with non-symmetrical constrained control. IEEE Trans. Aut. Control **38**(10), 1556–1560 (1993)
2. A. Benzaouia, Resolution of equation XA + XBX = HX and the pole assignment problem. IEEE Trans. Aut. Control **39**(10) (1994)
3. A. Benzaouia, A. Baddou, Piecewise linear constrained control for continuous-time systems. IEEE Trans. Automat. Control **44**(7), 1477–1481 (1999)
4. A. Benzaouia, M. Darouach, A. Hmamed, Solution of equation XA + XBX = HXE and the pole assignment for singular systems: Application to constrained control. IMA J. Math. Cont. Autom. **29**(3), 343–356 (2012)
5. L. Dai, Observers for discrete singular systems. IEEE Trans. Aut. Control **33**(2), 187–191 (1988)
6. L. Dai, *Singular Control Systems*, Lecture Notes in Control and Information Sciences (Springer, Berlin, 1989)
7. M. Darouach, M. Boutayeb, Design of observes for descriptor systems. IEEE Trans. Aut. Control **40**(7), 1323–1327 (1995)
8. M. Darouach, M. Zasadzinski, M. Hayar, Reduced-order observer design for descriptor systems with unknown inputs. IEEE Trans. Aut. Control **41**(7), 1068–1072 (1996)
9. M. Darouach, Solution to Sylvester equation associated to linear descriptor systems. Syst. Control Lett. **55**(9), 835–838 (2006)
10. B. Marx, D. Koenig, J. Ragot, Design of observers for Takagi-Sugeno descriptor systems with unknown inputs. IET Control Theory Appl. **1**(5), 1487–1495 (2007)
11. F. Mesquine, D. Mehdi, Constrained observer for linear continuous time systems. Int. J. Syst. Sci. **27**(12), 1363–1369 (1996)
12. F. Mesquine, D. Mehdi, A. Benzaouia, Constrained based full and minimal order observer control for linear systems, in *6th CMMNI, Tunis* (1998)
13. F. Mesquine, F. Tadeo, A. Benzaouia, Regulator problem for linear systems with constraints on control and its increment or rate. Automatica **40**(8), 1387–1395 (2004)

14. F. Mesquine, F. Tadeo, A. Benlamkadem, Constrained regulator problem for linear uncertain systems: control of a PH process. Math. Probl. Eng. (2006). doi:10.1155/MPE/2006/51874
15. J. Oreilly, *Observer for Linear Systems.* (Academic-press, London, 1983). Contributed Works
16. A.M. Perdon, M. Anderlucci, An unknown input observer for singular time-delay systems, in *14th Mediterranean Conference on Control and Automation, June 28–30, Ancona, Italy* (2006)
17. W.E. Schmitendorf, Design methodolgy for robust stabilizing controllers. J. Guid. **10**(3), 250–254 (1987)

Chapter 6
Constrained Control and Rate or Increment: An LMI Approach

6.1 Introduction

This chapter studies the stability of linear systems with input saturation and increment constraints. As seen before, apart from saturation constraints, different type of constraints, namely incremental or rate constraints, were introduced while considering practical applications. Symmetrical constraints on the input and its increment were considered for example in [7, 11, 14]. Hence, and as an extension to the non-symmetrical case, the regulator problem for linear continuous-time and discrete-time systems with input saturation and asymmetric constraints on its increment or rate in terms of an LMI problem is addressed. It is based on the previous results of [2, 8, 9], where the constraints are symmetric. Necessary and sufficient conditions of positive invariance for incremental domains with respect to autonomous systems are given. In this framework, a pole assignment procedure is linked to these conditions to design stabilizing controllers by state feedback. The resulting control respects control constraints and control increment or rate constraints also. As pointed out above, this chapter will also focus on the same problem of saturation on the control and asymmetric constraints on its increment or rate but in the context of LMI presented firstly in [3]. The first contribution of this chapter is to show that the dimension of the LMI proposed in [8, 9] can be reduced considerably by a resolution in a reduced-order state space. Further, the second contribution is to formulate under LMI form the results of [12, 13] that can handle asymmetric constraints on the increments.

6.2 Problem Presentation

The study thereafter is devoted to linear systems, as presented by Fig. 6.1, described by (6.1):

$$\delta x(t) = Ax(t) + Bsat(u(t)), \tag{6.1}$$

A. Benzaouia et al., *Saturated Control of Linear Systems*, Studies in Systems, Decision and Control 124, https://doi.org/10.1007/978-3-319-65990-9_6

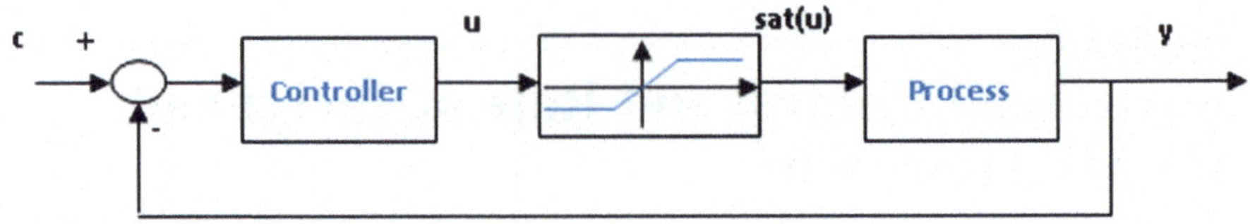

Fig. 6.1 Schema block of the studied system

where the operator δ is defined here as follows:

$$\delta x(t) = \begin{cases} x(t+1) & \text{for discrete-time case,} \\ \dot{x}(t) & \text{for continuous-time case,} \end{cases}$$

vector $x \in \mathbb{R}^n$ is the state and $u \in \mathbb{R}^m$ is the constrained control. The saturation function is assumed here to be symmetric and normalized:

$$sat(u_j) = \begin{cases} 1 & \text{if } u_j \geq 1 \\ u_j & \text{if } -1 < u_j < 1 \\ -1 & \text{if } u_j \leq -1 \end{cases}, \quad j = 1, \ldots, m.$$

A and B are constant matrices of appropriate size and satisfy the following assumptions:

- (AS1): The pair (A, B) is controllable,
- (AS2): The open-loop system has m unstable or undesirable eigenvalues.

Let the increment or the rate of the control be asymmetrically constrained as follows:

$$\Delta u \in \Omega_\Delta \subset \mathbb{R}^m, \tag{6.2}$$

where the operator Δ is defined as follows:

$$\Delta u(t) = \begin{cases} u(t+1) - u(t) & \text{for discrete-time case,} \\ \dot{u}(t) & \text{for continuous-time case,} \end{cases}$$

and Ω_Δ is the set of admissible control increments or rate defined as:

$$\Omega_\Delta = \{u \in \mathbb{R}^m / -q_2 \leq \Delta u \leq q_1;\ q_1,\ q_2 \in \mathbb{R}_+^m\}. \tag{6.3}$$

The set Ω_Δ is a non-symmetrical polyhedral set as is generally the case in practical situations. Further, the notation q will stand for:

$$q = \begin{bmatrix} q_1 \\ q_2 \end{bmatrix}. \tag{6.4}$$

In what follows, we are interested on the synthesis of stabilizing controllers for this class of linear systems using a state feedback control law:

$$u(t) = Fx(t), \tag{6.5}$$

which writes the closed-loop system as,

$$\delta x(t) = Ax(t) + Bsat(Fx(t)), \tag{6.6}$$

subject to input saturation and constraints on its increment or rate. The main objective of this chapter is to design stabilizing controllers by saturating state feedback control while the constraints on its increment or rate are always respected in the framework of LMI. The results presented in the sequel are based on the projection technique presented thereafter.

6.3 Projection Technique

The projection technique concerns the resolution of the algebraic equation:

$$FA + FBF = HF, \tag{6.7}$$

by using the LMI technique.

Consider the following transformation of the system:

$$x = Qz = [Q_o \,|\, Q_m] \begin{bmatrix} z_o \\ z_m \end{bmatrix}; \; z_o \in \mathbb{R}^{n-m}, \; z_m \in \mathbb{R}^m, \tag{6.8}$$

where the matrix $Q \in \mathbb{R}^{n \times n}$ is orthonormal and is partitioned as follows:

$$Q = [Q_o \mid Q_m], \; Q_o \in \mathbb{R}^{n \times (n-m)}, \; Q_m \in \mathbb{R}^{n \times m},$$

with

$$Q^T Q = QQ^T = \mathbb{I}_n, \; Q_m^T Q_m = \mathbb{I}_m, \; Q_m^T Q_o = \odot, \tag{6.9}$$

where the columns of Q_o span $S_o \triangleq Ker(F)$ and the columns of Q_m span its complement S_m in $\mathbb{R}^n$.

Matrix Q can be obtained from a Schur decomposition of matrix A by reordering, if necessary, its Schur blocks [6]. In the orthonormal basis formed by the columns of matrix Q, the open-loop system (6.1) is represented by:

$$\begin{bmatrix} \delta z_o(t) \\ \delta z_m(t) \end{bmatrix} = A_Q \begin{bmatrix} z_o(t) \\ z_m(t) \end{bmatrix} + B_Q sat(u(t)), \tag{6.10}$$

where

$$A_Q = Q^T A Q = \begin{bmatrix} A_o & A_2 \\ \odot & A_m \end{bmatrix}, A_o \in \mathbb{R}^{(n-m)\times(n-m)}, A_m \in \mathbb{R}^{m\times m},$$
$$B_Q = \begin{bmatrix} Q_o^T \\ Q_m^T \end{bmatrix} B = \begin{bmatrix} B_o \\ B_m \end{bmatrix}, B_o \in \mathbb{R}^{(n-m)\times m}, B_m \in \mathbb{R}^{m\times m}. \tag{6.11}$$

According to assumption $(AS2)$ and the reordering possibility of the Schur blocks of matrix A, the obtained matrix A_o is Hurwitz, that is, there always exists a positive definite matrix: P_o such that

$$\begin{cases} A_o^T P_o A_o - P_o \prec 0 \text{ for discrete-time case,} \\ A_o^T P_o + P_o A_o \prec 0 \text{ for continuous-time case.} \end{cases} \tag{6.12}$$

By virtue of this transformation, the control becomes:

$$u(t) = Fx(t) = FQz(t) = [FQ_o \;\; FQ_m]z(t). \tag{6.13}$$

If one takes $F = F_m Q_m^T$, $F_m \in \mathbb{R}^{m\times m}$, then, according to (6.9), it is possible to deduce that:

$$u(t) = [F_m Q_m^T Q_o \;\; F_m Q_m^T Q_m]z(t) = [\odot \;\; F_m]z(t) = F_m z_m(t). \tag{6.14}$$

This projection method leads in fact to obtain two reduced-order systems given by:

$$\delta z_m(t) = A_m z_m(t) + B_m sat(F_m z_m(t)), \tag{6.15}$$
$$\delta z_o(t) = A_o z_o(t) + c(t), \tag{6.16}$$
$$c(t) = A_2 z_m(t) + B_o sat(F_m z_m(t)),$$

where the undesirable m eigenvalues of the spectrum of matrix A are isolated in the spectrum of matrix A_m. The spectrum of matrix A_o contains the $n-m$ stable eigenvalues of the open-loop system according to assumption $(AS2)$. The algebraic equation (6.7) is also transformed to the following:

$$F_m A_m + F_m B_m F_m = H F_m. \tag{6.17}$$

Theorem 6.1 [2] *A matrix F of full rank is the unique solution of the Eq. (6.7), where matrices A, B satisfy assumptions $(AS1) - (AS2)$ and $H \in \mathbb{R}^{m\times m}$ a given matrix, if and only if there exist non-singular matrices $X \in \mathbb{R}^{m\times m}$ and $Y \in \mathbb{R}^{m\times m}$ solutions of the following algebraic equations:*

$$\begin{cases} A_m X + B_m Y - X J = 0, \\ HY - YJ = 0, \end{cases} \tag{6.18}$$

where the matrix J denotes the Jordan form of matrix H. Moreover, the unique solution of (6.7) is $F = F_m Q_m^T = Y X^{-1} Q_m^T$ and matrices A_m, B_m, Q_m are given by (6.9) and (6.11).

Note that $F_m = YX^{-1}$ is in fact the unique solution of the reduced order Eq. (6.17).

Proof

Necessity: Let matrix F with full rank be the unique solution of the algebraic equation (6.7). Using the transformation (6.11), Eq. (6.7) becomes:

$$FQA_QQ^T + F[Q_o \;\; Q_m]B_QF = HF \tag{6.19}$$

The following developments can be obtained:

$$\begin{aligned} F[Q_o \;\; Q_m]A_Q\begin{bmatrix} Q_o^T \\ Q_m^T \end{bmatrix} + [0 \;\; FQ_m]\begin{bmatrix} B_o \\ B_m \end{bmatrix}F &= HF, \\ [0 \;\; FQ_m]\begin{bmatrix} A_o & A_2 \\ 0 & A_m \end{bmatrix}\begin{bmatrix} Q_o^T \\ Q_m^T \end{bmatrix} + FQ_mB_mF &= HF, \\ [0 \;\; F_m]\begin{bmatrix} A_o & A_2 \\ 0 & A_m \end{bmatrix}\begin{bmatrix} Q_o^T \\ Q_m^T \end{bmatrix} + F_mB_mF_mQ_m^T &= HF_mQ_m^T, \\ F_mA_mQ_m^T + F_mB_mF_mQ_m^T &= HF_mQ_m^T. \end{aligned} \tag{6.20}$$

Multiplying this equality by Q_m at the right, the reduced algebraic equation (6.17) is obtained. Using the same transformations (6.11), matrix F_m is the non-singular unique solution of the reduced algebraic equation (6.17). According to [1], this solution is given by $F_m = YX^{-1}$ where X and Y satisfy:

$$F_mX = Y, \tag{6.21}$$

$$HY = YJ. \tag{6.22}$$

Multiply in the right Eq. (6.17) by matrix X, it follows,

$$F_m(A_mX + B_mF_mX) = HF_mX. \tag{6.23}$$

Using (6.21) and the fact that matrix F_m is non-singular, then,

$$A_mX + B_mX = XY^{-1}HYX. \tag{6.24}$$

Using (6.22), one obtains the Sylvester equation (6.18). In conclusion, the unique solution of full rank of the algebraic equation (6.17) is given by $F = YX^{-1}Q_m^T$ with matrices X and Y are solutions of the system (6.18).

Sufficiency: Let X and Y be the non-singular solutions of the system (6.18). According to the resolution of the Sylvester equation (6.17), one can write:

$$(A_m + B_m F_m)X = XJ, \tag{6.25}$$

which means that the columns of matrix X, which is non-singular, presents the eigenvectors of the reduced matrix in closed-loop associated to the eigenvalues of matrix H. It follows that:

$$\begin{aligned}(A_m + B_m F_m) &= XJX^{-1}, \\ F_m &= YX^{-1}.\end{aligned} \tag{6.26}$$

Following the same development as in the necessity proof from (6.20) to (6.19), one can obtain Eq. (6.7). Finally, one can note that matrix F_m is also non-singular, that is, matrix F is of full rank. □

6.4 Constrained Control Synthesis

This section presents the asymptotic stability condition for linear systems (6.6) with constraints on the control increment or rate by means of LMI technique. The synthesis design of adequate controllers in a state space of reduced order is also presented under LMI form. In addition, the relaxation of such LMI is achieved by introducing an uncertainty term in the algebraic equation (6.7). First, the stability of the saturated system in closed loop (6.6) without constraints on the increment or rate of the control is in concern. For this, define the following subsets of $\mathbb{R}^n$:

$$\varepsilon(P, \rho) = \{x \in \mathbb{R}^n / x^T P x \leq \rho, \rho > 0\}, \tag{6.27}$$

$$\mathscr{L}(G) = \{x \in \mathbb{R}^n / |G_j x| \leq 1,\ j = 1, \ldots, m\}, \tag{6.28}$$

with $P = P^T$ a positive definite matrix and $G \in \mathbb{R}^{m \times n}$. Thus, $\varepsilon(P, \rho)$ is an ellipsoid, while $\mathscr{L}(G)$ is a polyhedral. One can also define the set $\mathscr{L}(F)$ consisting of states for which the saturation does not occur. Recall that η as 2^m and $e = [1 \quad \cdots \quad 1]^T \in \mathbb{R}^m$.

Theorem 6.2 *Given a positive scalar ρ, if there exist a positive definite matrix $P = P^T \in \mathbb{R}^{n \times n}$, a matrix $G \in \mathbb{R}^{m \times n}$, and η matrices $H_i \in \mathbb{R}^{m \times m}$ such that $\forall i \in [1, \eta]$,*

$$\begin{cases} Ac_i^T P Ac_i - P \prec 0 \ \textit{for discrete-time case}, \\ Ac_i^T P + P Ac_i \prec 0 \ \textit{for continuous-time case}, \end{cases} \tag{6.29}$$

$$FA + FB(D_i F + D_i^- G) = H_i F, \tag{6.30}$$

$$\begin{cases} \widetilde{(H_i)_d} \;\; \rho \le q \textit{ for discrete-time case}, \\ \widetilde{(H_i)_c} \;\; \rho \le q \textit{ for continuous-time case}, \end{cases} \tag{6.31}$$

$$\varepsilon(P, \rho) \subset \mathscr{L}(G), \tag{6.32}$$

where $Ac_i = A + B(D_i F + D_i^- G)$, $\rho^T = [e^T e^T]$ *and vector* q *is given by (6.4), then, the system in closed loop (6.6) is asymptotically stable at the origin while the constraints on the control increment or rate are respected for all* $x_o \in \varepsilon(P, \rho)$.

Proof If there exist matrices $P = P^T$ and G satisfying condition (6.32), then, by virtue of Lemma 1.3 [5], there exists scalars γ_i, $i = 1, \ldots, \eta$ such that:

$$sat(Fx(t)) = \sum_{i=1}^{\eta} \gamma_i(t)(D_i F + D_i^- G)x(t),$$

$$\gamma_i(t) \ge 0; \;\; \sum_{i=1}^{\eta} \gamma_i(t) = 1, \;\; \forall t. \tag{6.33}$$

This allows to rewrite system (6.6) as follows:

$$\delta x(t) = \sum_{i=1}^{\eta} \gamma_i(t)[A + B(D_i F + D_i^- G)]x(t). \tag{6.34}$$

According to Theorem 1.10, condition (6.29) guarantees the asymptotic stability of the saturated system in closed loop. Now, one has to prove that the constraints on the control increment or rate are respected. For this and from the feedback control law (6.5), the control dynamic is given by:

$$\delta u(t) = F\delta x(t) = \sum_{i=1}^{\eta} \gamma_i(t) F[A + B(D_i F + D_i^- G)]x(t). \tag{6.35}$$

Taking account of Eq. (6.30), the last can be rewritten as:

$$\delta u(t) = \sum_{i=1}^{\eta} \gamma_i(t) H_i u(t). \tag{6.36}$$

Recall that this system has the following constraints:

$$-e \le u \le e, \tag{6.37}$$

$$-q_2 \le \Delta u \le q_1. \tag{6.38}$$

Keeping in mind that the scalars γ_i satisfy (6.33), one can easily apply the results of [12] which are satisfied with conditions (6.31) and by noting that

$$\widetilde{(\sum_{i=1}^{\eta} \gamma_i(t) H_i)_c} \leq \sum_{i=1}^{\eta} \gamma_i(t) \widetilde{(H_i)_c}, \tag{6.39}$$

conclude that the constraints on the control increment or rate are also respected. □

Remark 6.1
- Conditions (6.29) and (6.30) are not completely independent. Equation (6.30) is in fact a pole assignment equation which realizes $\sigma(Ac_i) = \Lambda_o \bigcup \sigma(H_i)$, where Λ_o contains the $n - m$ stable eigenvalues of matrix A according to assumption $(AS2)$. In this sense, if matrices H_i are not Hurwitz, condition (6.29) will never be satisfied. Furthermore, based on the proof of [12] concerning inequalities (6.31), it is always possible to consider a null component of the constraints on the control increment or rate q.
- The results of [12] constitute a particular case of conditions given by (6.30)–(6.31) of Theorem 6.2 with $F = G$. Hence, the region of asymptotic stability $\mathscr{L}(F)$, obtained with [12] and where the saturation are not allowed, is limited to the region of linear behavior. Further, the region of asymptotic stability, obtained here, tolerates the saturation to take effect and is contained in $\mathscr{L}(G)$.

The results of Theorem 6.2 can be developed to obtain LMI's conditions which lead to the adequate controller design. In this case, Theorem 6.2 is applied to the reduced-order system with matrices A_m and B_m. In this sense, conditions (6.29) and (6.32) are transformed to LMI according to [8, 9], while the algebraic equations (6.30) are transformed to LMI by virtue of Theorem 6.1. Lemma below applies conditions (6.29) and (6.32) to the reduced-order system but without constraints on the control increment. The last will be considered later on.

Lemma 6.1 *For a given positive scalar ρ, if there exist matrices $Y \in \mathbb{R}^{m\times m}$ and $Z \in \mathbb{R}^{m\times m}$ and a positive definite matrix $X = X^T \in \mathbb{R}^{m\times m}$, solutions of the following LMI:*

$$\begin{cases} \begin{bmatrix} X & [A_m X + B_m(D_i Y + D_i^- Z)]^T \\ * & X \end{bmatrix} \succ 0, & \textit{for discrete-time case} \\ \left[A_m X + B_m(D_i Y + D_i^- Z)\right]^{sym} \prec 0, & \textit{for continuous-time case,} \end{cases} \tag{6.40}$$

$$\begin{bmatrix} 1/\rho & Z_j \\ * & X \end{bmatrix} \succ 0, i = 1, \ldots, \eta; \quad j = 1, \ldots, m, \tag{6.41}$$

then the saturated system without constraints on the increment or rate in closed loop (6.6) is asymptotically stable at the origin $\forall x_o \in \varepsilon(P, \rho)$ with

$$F = YX^{-1}Q_m^T, \tag{6.42}$$
$$G = ZX^{-1}Q_m^T, \tag{6.43}$$
$$P_m = X^{-1}, \tag{6.44}$$
$$P = Q\begin{bmatrix} P_o & \odot \\ \odot & P_m \end{bmatrix}Q^T, \tag{6.45}$$

where matrices Q_m, Q, P_o are given by (6.9) and (6.12), respectively.

Proof Let the assumption $(AS2)$ be held, the Schur decomposition of the system matrices (A, B) is used by reordering, if necessary, its Schur blocks. According to the transformation Q, system (6.34) is rewritten with the following equation,

$$\delta z(t) = \sum_{i=1}^{\eta} \gamma_i(t)[A_Q + B_Q(D_i F_Q + D_i^- G_Q)]z(t), \tag{6.46}$$

where matrices A_Q and B_Q are given by (6.11) and $F_Q = FQ$, $G_Q = GQ$. One can rewrite matrices F_Q and G_Q as follows:

$$\begin{aligned} F_Q &= FQ = [FQ_o \;\; FQ_m], \\ G_Q &= GQ = [GQ_o \;\; GQ_m]. \end{aligned} \tag{6.47}$$

Taking into account (6.42) and (6.43), $F = F_m Q_m^T$, $G = G_m Q_m^T$ with $F_m = YX^{-1}$ and by virtue of (6.9), the transformation Q leads to

$$F_Q = [F_m Q_m^T Q_o \;\; F_m Q_m^T Q_m] = [\odot \;\; F_m], \tag{6.48}$$
$$G_Q = [G_m Q_m^T Q_o \;\; G_m Q_m^T Q_m] = [\odot \;\; G_m]. \tag{6.49}$$

Thus, system (6.46) can be partitioned into the two following subsystems of dimensions $n - m$ and m, respectively,

$$\begin{aligned} \delta z_o(t) &= A_o z_o(t) + c(t), \\ c(t) &= \sum_{i=1}^{\eta} \gamma_i(t)[A_2 + B_o(D_i F_m + D_i^- G_m)]z_m(t), \end{aligned} \tag{6.50}$$

and

$$\delta z_m(t) = \sum_{i=1}^{\eta} \gamma_i(t)[A_m + B_m(D_i F_m + D_i^- G_m)]z_m(t). \tag{6.51}$$

The result of Theorem 1.10 is then applied to the reduced-order system (6.51) leading to the following conditions of stability:

$$\begin{cases} \Phi_i^T P_m \Phi_i - P_m \prec 0 & \text{for discrete-time case} \\ \Phi_i^T P_m + P_m \Phi_i \prec 0 & \text{for continuous-time case} \end{cases} \quad i = 1, \ldots, \eta, \tag{6.52}$$

where $\Phi_i = A_m + B_m(D_i F_m + D_i^- G_m)$. In this case, using the change of variables $F_m = YX^{-1}$, $G_m = ZX^{-1}$, and $P_m = X^{-1}$, conditions (6.40) guarantee the asymptotic stability of the reduced-order system (6.51). LMIs (6.40) are easily obtained from (6.52) by using the Schur complement for the discrete-time case and directly for the continuous-time case [8, 9]. Consequently, function $V_m(z_m) = z_m^T P_m z_m$ is a Lyapunov function of the reduced-order system (6.51). Recall that the spectrum of matrix A_o of the subsystem (6.50) contains the $n - m$ stable eigenvalues of the open-loop system according to assumption $(AS2)$. That is, there always exists a positive definite matrix $P_o = P_o^T$ such that inequality (6.12) is satisfied. Note that $c(t)$ is considered as a bounded disturbance such that $lim_{t\to\infty} c(t) = 0$ by virtue of the asymptotic stability of the reduced-order system. Then, the asymptotic stability of system (6.50) is realized. Consequently, the obtained system (6.46) after using the transformation Q is also asymptotically stable, and the corresponding Lyapunov function is given by $V_z(z) = z^T Rz$, with:

$$R = \begin{bmatrix} P_o & \odot \\ \odot & P_m \end{bmatrix}. \tag{6.53}$$

Recall that $z = Q^T x$; this leads to $V_z(z) = z^T Rz = x^T(QRQ^T)x = V(x)$. Then, function $V(x) = x^T Px$ is also a Lyapunov function of the saturated system in closed loop (6.6) written equivalently as (6.34) with matrix $P = QRQ^T$ which is positive definite. Furthermore, the results in [4] can be used to satisfy the set inclusion $\varepsilon(P_m, \rho) \subset \mathscr{L}(G_m)$ under the LMI (6.41).

In order to complete the proof, the following implication has to be shown,

$$\varepsilon(P_m, \rho) \subset \mathscr{L}(G_m) \text{ implies } \varepsilon(P, \rho) \subset \mathscr{L}(G).$$

For this, expression (6.49) is used to show the inclusion $\varepsilon(R, \rho) \subset \mathscr{L}(G_Q)$. According to [4], this inclusion is equivalent to $\rho(G_Q)_j R^{-1}(G_Q)_j^T \leq 1, j = 1, \ldots, m$. Using expressions (6.53) and (6.49), one can easily obtain:

$$\begin{aligned} \rho(G_Q)_j R^{-1}(G_Q)_j^T &= \rho[\odot \ (G_m)_j] \begin{bmatrix} P_o^{-1} & \odot \\ \odot & P_m^{-1} \end{bmatrix} \begin{bmatrix} \odot \\ (G_m)_j^T \end{bmatrix} \\ &= \rho(G_m)_j P_m^{-1}(G_m)_j^T, \ j = 1, \ldots, m. \end{aligned} \tag{6.54}$$

This is equivalent to have

$$\varepsilon(P_m, \rho) \subset \mathscr{L}(G_m) \text{ iff } \varepsilon(R, \rho) \subset \mathscr{L}(G_Q).$$

Now, rewrite $\rho(G_Q)_j R^{-1}(G_Q)_j^T$ by using $G_Q = GQ$:

$$\begin{aligned}\rho(G_Q)_j R^{-1}(G_Q)_j^T &= \rho(GQ)_j R^{-1}(GQ)_j^T\\ &= \rho G_j(QR^{-1}Q^T)G_j^T\\ &= \rho G_j P^{-1}G_j^T,\ j=1,\dots,m.\end{aligned}$$

That is,

$$\varepsilon(R,\rho)\subset\mathscr{L}(G_Q)\ \text{iff}\ \varepsilon(P,\rho)\subset\mathscr{L}(G).$$

Now, we have to show that $\varepsilon(P_m,\rho)\subset\mathscr{L}(G_m)$. LMI (6.41) is equivalent to the following developments:

$$\begin{aligned}\rho Z_j X^{-1}Z_j^T &< 1\\ \rho(Z_jX^{-1})X(Z_jX^{-1})^T &< 1\\ \rho(ZX^{-1})_jX(ZX^{-1})_j^T &< 1\end{aligned}$$

using $ZX^{-1}=G_m$, $X=P_m^{-1}$, one obtains:

$$\rho G_{mj}P_m^{-1}G_{mj}^T < 1, \tag{6.55}$$

hence, $\varepsilon(P_m,\rho)\subset\mathscr{L}(G_m)$.
Finally, the inclusion $\varepsilon(P,\rho)\subset\mathscr{L}(G)$ is ensured by the reduced-order LMI (6.41), which ends the proof. □

The constraints on the increment or rate can be handled using lemma above. The controller design for the linear system with saturation and constraints on the control increment or rate according to Theorem 6.2 is now given.

Let $W=W^T\succ 0$ and η matrices H_i be chosen, for all $i=1,\dots,\eta$ such that

$$\sigma(A)\bigcap\sigma(H_i)=\emptyset, \tag{6.56}$$

$$\begin{cases} J_i^TWJ_i-W\prec 0, & \text{for discrete-time case,}\\ J_i^TW+WJ_i\prec 0, & \text{for continuous-time case,}\end{cases} \tag{6.57}$$

where matrix J_i represents the Jordan form of matrix H_i.

Lemma 6.2 *For a given positive scalar ρ and η diagonalizable matrices $H_i\in\mathbb{R}^{m\times m}$ satisfying (6.56), (6.57), and (6.31), if there exist a positive definite matrix $X=X^T\in\mathbb{R}^{m\times m}$, non-singular matrices $Y\in\mathbb{R}^{m\times m}$, and $Z\in\mathbb{R}^{m\times m}$ solutions of the following LMI:*

$$\begin{cases}\begin{bmatrix} X & J_i^TX\\ * & X\end{bmatrix}\succ 0, & \textit{for discrete-time case,}\\ XJ_i+J_i^TX\prec 0, & \textit{for continuous-time Case,}\end{cases} \tag{6.58}$$

$$A_mX+B_m(D_iY+D_i^-Z)-XJ_i=0, \tag{6.59}$$

$$H_iY-YJ_i=0, \tag{6.60}$$

$$\begin{bmatrix} 1/\rho & Z_j \\ * & X \end{bmatrix} \succ 0, \; i = 1, \ldots, \eta; \; j = 1, \ldots, m,$$

then the saturated system with asymmetric constraints on the increment or rate in closed loop (6.6) is asymptotically stable at the origin $\forall x_o \in \varepsilon(P, \rho)$, *with*

$$F = YX^{-1}Q_m^T, \tag{6.61}$$

$$G = ZX^{-1}Q_m^T, \tag{6.62}$$

$$P = Q\begin{bmatrix} P_o & \odot \\ \odot & X^{-1} \end{bmatrix}Q^T, \tag{6.63}$$

where matrices Q_m, Q, P_o *are given by (6.9) and (6.12), respectively.*

Proof The projection technique presented previously can be applied to the system (6.6) leading to the reduced-order system (6.51). The use of the change of variables (6.61)–(6.62) allows the extension of the results of Theorem 6.1 to rewrite equivalently Eq. (6.30) under the LMI form of (6.59)–(6.60). For this, multiply Eq. (6.59) on the left by F_m and on the right by X^{-1}, successively. It follows that:

$$F_m A_m + F_m B_m(D_i Y X^{-1} + D_i^- Z X^{-1}) - F_m X J_i X^{-1} = 0, \quad i = 1, \ldots, \eta \tag{6.64}$$

Noting that $F_m = YX^{-1}, G_m = ZX^{-1}$ and taking account of Eq. (6.60), the nonlinear algebraic equation associated to the the reduced-order system (6.51) is then obtained:

$$F_m A_m + F_m B_m(D_i F_m + D_i^- G_m) = H_i F_m, \quad i = 1, \ldots, \eta. \tag{6.65}$$

Note that the reciprocal is also true if matrices Y and Z are non-singular. The problem now is to show that if Eq. (6.65) is satisfied for the reduced-order system (6.51), then the Eq. (6.30) is also satisfied for system (6.34). Using expressions of matrices F_Q and G_Q given by (6.48)–(6.49) and Eq. (6.65), matrices A_Q and B_Q being given by (6.11), one can write

$$\begin{aligned} F_Q A_Q + F_Q B_Q(D_i F_Q + D_i^- G_Q) &= [\odot \quad F_m A_m] + [\odot \quad F_m B_m(D_i F_m + D_i^{-1} G_m)] \\ &= [\odot \quad H_i F_m] \\ &= H_i F_Q. \end{aligned} \tag{6.66}$$

Equation (6.66) can also be written as follows:

$$FQQ^T AQ + FQQ^T B(D_i FQ + D_i^- GQ) = H_i FQ. \tag{6.67}$$

By multiplying on the right Eq. (6.67) by matrix Q^T, and using the property of the Schur matrix $QQ^T = \mathbb{I}$, Eq. (6.30) follows. However, it is useful to note that Eq. (6.30), even satisfied, is not really used with this projection technique. One can only compute directly the control dynamics by using the reduced-order system (6.51) and the associated Eq. (6.65), leading to the following developments:

$$
\begin{aligned}
\delta u(t) &= F_m \delta z_m(t) \\
&= \sum_{i=1}^{\eta} \gamma_i(t) F_m [A_m + B_m(D_i F_m + D_i^- G_m)] z_m(t) \\
&= \sum_{i=1}^{\eta} \gamma_i(t) H_i u(t).
\end{aligned} \tag{6.68}
$$

Note that matrices H_i are chosen diagonalizable and satisfying (6.56)–(6.57) in order to ensure that the solution Y is non-singular, there exists a solution of (6.58) and the pole assignment of Eq. (6.59) is feasible. In addition to this, matrices H_i are also chosen according to conditions (6.31), hence the constraints on the control increment or rate are respected.

Further, for the stability purpose of the system (6.6), Lemma 6.1 is used to transform inequalities (6.29) of Theorem 6.2 to obtain the following LMI:

$$
\begin{cases}
\begin{bmatrix} X & [A_m X + B_m(D_i Y + D_i^- Z)]^T \\ * & X \end{bmatrix} \succ 0, \text{ for discrete-time case} \\
\left[A_m X + B_m(D_i Y + D_i^- Z)\right]^{sym} \prec 0, \text{ for continuous-time case.}
\end{cases} \tag{6.69}
$$

However, with this projection technique, these inequalities can also be transformed. In fact, if equalities (6.59)–(6.60) are satisfied, one can substitute (6.59) into (6.69) to obtain (6.58). □

Comments 6.1

- *The results of Lemmas 6.1 and 6.2 propose two main ideas: the first is to allow the integration of the resolution of the algebraic equation (6.30) among the LMIs of Lemma 6.2 by means of LMIs (6.59)–(6.60); the second, which is of great interest and presented for the first time in [3], is to reduce the dimension of the LMIs (6.29) of [8, 10], allowing an important computing time economy.*
- *To obtain matrices H_i satisfying conditions (6.31), one can use the linear programming technique proposed in [12].*
- *Compared to the results of [12], one can note that the results given above present a different way to deal with the studied problem. Further, the obtained results are given under LMI formulation. Furthermore, the projection technique as proposed above is also a new tool to reduce the computation load.*

In order to present a different way to design the required controllers of Lemma 6.2 by relaxing equalities (6.59)–(6.60), one can use the following approach. Let the reduced-order algebraic equations (6.65) be satisfied only with an uncertainty term as follows,

$$
F_m A_m + F_m B_m(D_i F_m + D_i^- G_m) - H_i F_m = M_i, \; i = 1, \ldots, \eta, \tag{6.70}
$$

hence, the control dynamic becomes:

$$\delta u(t) = \sum_{i=1}^{\eta} \gamma_i(t)(H_i F_m + M_i) z_m(t). \tag{6.71}$$

If one compute matrix G_m as a non-singular one, an uncertain system is obtained and it is given by,

$$\delta u(t) = \sum_{i=1}^{\eta} \gamma_i(t) H_i u(t) + \sum_{i=1}^{\eta} \gamma_i(t) L_i p(t), \tag{6.72}$$
$$L_i = M_i G_m^{-1};\ p(t) = G_m z_m(t),$$

where $p(t)$ is considered as a disturbance entry. Further, if $z_m(t) \in \mathscr{L}(G_m)$, $\forall t$, one should have,

$$-e \le p(t) \le e, \quad \forall t. \tag{6.73}$$

Assume now that the tilde transform of the uncertainty terms $\widetilde{(L_i)_d}$ is bounded by a known positive matrix $\Gamma \in \mathbb{R}^{2m\times 2m}$,

$$\widetilde{(L_i)_d} \le \Gamma, i = 1, \ldots, \eta. \tag{6.74}$$

According to [13] for the continuous-time case (while the discrete-time case is obtained in a similar way), and using the same arguments on the scalars γ_i, conditions (6.31) are to be changed to the following:

$$\begin{cases} \widetilde{(H_i)_d}\ \rho + \Gamma\rho \le q & \text{for discrete-time case,} \\ \widetilde{(H_i)_c}\ \rho + \Gamma\rho \le q & \text{for continuous-time case.} \end{cases} \tag{6.75}$$

Theorem 6.3 *For given positive scalar ρ, positive matrix $\Gamma \in \mathbb{R}^{2m\times 2m}$ and η Hurwitz matrices $H_i \in \mathbb{R}^{m\times m}$ satisfying (6.75), if there exist a positive definite matrix $X = X^T \in \mathbb{R}^{m\times m}$, matrices $Y \in \mathbb{R}^{m\times m}$, and $Z \in \mathbb{R}^{m\times m}$ solutions of the following LMIs:*

$$\begin{cases} \begin{bmatrix} X & [A_m X + B_m(D_i Y + D_i^- Z)]^T \\ * & X \end{bmatrix} \succ 0, & \textit{for discrete-time case.} \\ \left[A_m X + B_m(D_i Y + D_i^- Z)\right]^{sym} \prec 0, & \textit{for continuous-time case.} \end{cases} \tag{6.76}$$

$$\begin{bmatrix} 1/\rho & Z_j \\ * & X \end{bmatrix} \succ 0, \tag{6.77}$$

$$Z + Z^T \succ 0, \tag{6.78}$$

$$i = 1, \ldots, \eta;\ j = 1, \ldots, m,$$

such that inequalities (6.74) are satisfied with $F_m = YX^{-1}, G_m = ZX^{-1}$ *and* M_i *given by (6.70), then the saturated system with asymmetric constraints on the increment or rate in closed loop (6.6) is asymptotically stable at the origin* $\forall x_o \in \varepsilon(P, \rho)$, *with*

$$F = F_m Q_m^T, \tag{6.79}$$

$$G = G_m Q_m^T, \tag{6.80}$$

$$P_m = X^{-1}, \tag{6.81}$$

$$P = Q \begin{bmatrix} P_o & \odot \\ \odot & P_m \end{bmatrix} Q^T, \tag{6.82}$$

where matrices Q_m, Q, P_o *are given by (6.9) and (6.12), respectively.*

Proof The proof follows readily from the above development. Note that it is easy to require that matrix G_m be non-singular by adding the constraint (6.78). The proof is also based on the fact that $z_m(t) \in \mathcal{L}(G_m)$, $\forall t$. Thus, the following has to be shown,

$$x(t) \in \mathcal{L}(G) \text{ implies } z_m(t) \in \mathcal{L}(G_m), \forall t. \tag{6.83}$$

For this, as it is known that $x(t) \in \varepsilon(P, \rho) \subset \mathcal{L}(G)$, $\forall t$, because the set $\varepsilon(P, \rho)$ is a contractively invariant set, then, $\rho^{-1}x^T(t)Px(t) \leq 1, \forall t$. Using the transformation $x(t) = Qz(t)$, it follows that

$$\begin{aligned} \rho^{-1}z^T(t)(Q^T PQ)z(t) &= \rho^{-1}z^T(t)Rz(t) \leq 1 \\ &= \rho^{-1}z_o^T(t)P_o z_o(t) + \rho^{-1}z_m^T(t)P_m z_m(t) \leq 1, \forall t. \end{aligned}$$

This last inequality implies that $\rho^{-1}z_m^T(t)P_m z_m(t) < 1, \forall t$; that is, $z_m(t) \in \varepsilon(P_m, \rho)$, which according to (6.77) ensures that $z_m(t) \in \mathcal{L}(G_m), \quad \forall t.$ □

Remark 6.2 In order to satisfy condition (6.75), due to the block symmetry of matrix $\widetilde{(H_i)}_c$, one can choose the positive matrix Γ as follows,

$$\Gamma = \begin{bmatrix} \Gamma_1 & \Gamma_2 \\ \Gamma_2 & \Gamma_1 \end{bmatrix}.$$

In this case, with vector $\rho = [e^T \ e^T]^T$ one has to only check that for $i = 1, \ldots, \eta$,

$$\begin{cases} |H_i - \mathbb{I}_m|e + (\Gamma_1 + \Gamma_2)e \leq min(q_1, q_2) & \text{for discrete-time case,} \\ |H_i|e + (\Gamma_1 + \Gamma_2)e \leq min(q_1, q_2) & \text{for continuous-time case.} \end{cases} \tag{6.84}$$

One can also note that conditions (6.75) imply conditions (6.31).

Now, the steps of calculations followed during the development of this new approach by introducing the LMIs are summarized in the following algorithm. It is worth to recall that the use of a Schur decomposition guarantees numerical robustness in the computation of the open-loop eigenvalues, while it determines a new basis for the associated subspaces [6].

Algorithm 6.1

- Step 1: Check if matrix A satisfies assumption $(AS2)$, else, use the technique of augmentation and go to Step 4 with $A_m = A$, $B_m = B_a$ and $m = n$.
- Step 2: Apply a Schur decomposition to matrix A by reordering, if necessary, its Schur blocks to have matrix $Q_m \in R^{n\times m}$ and the reduced-order system (6.51) associated with the undesirable eigenvalues of matrix A.
- Step 3: Compute a positive definite matrix $P_o = P_o^T$ satisfying (6.12).
- Step 4: Give η diagonalizable matrices $H_i \in R^{m\times m}$ satisfying (6.56), (6.57), and (6.31).
- Step 5: If the LMIs (6.58)–(6.61) are feasible, continue with the solutions $X,\ Y,\ Z$, else, solve the LMIs (6.76)–(6.78) with an adequate choice of positive matrices $\Gamma_1 \in R^{m\times m}$ and $\Gamma_2 \in R^{m\times m}$ satisfying (6.74) and (6.75).
- Step 6: Compute the matrices $F = F_m Q_m^T = YX^{-1}Q_m^T, G = G_m Q_m^T = ZX^{-1}Q_m^T,$

$$P = Q\begin{bmatrix} P_o & \odot \\ \odot & X^{-1} \end{bmatrix} Q^T$$

Example 6.1 In order to illustrate the use of the proposed methodology, consider the same continuous-time system treated by [2] and the references therein:

$$A = \begin{bmatrix} -5 & 1 & 0 \\ 0 & 1 & 1 \\ 1 & 1 & 1 \end{bmatrix};\ B = \begin{bmatrix} 0 & 0 \\ 0 & 1 \\ 1 & 0 \end{bmatrix},\ \rho = 3.$$

Let the asymmetric constraints on the control rate be as follows:

$$q_1 = \begin{bmatrix} 15 \\ 16 \end{bmatrix};\ q_2 = \begin{bmatrix} 12.5 \\ 15 \end{bmatrix}.$$

The open-loop eigenvalues of the system are given by:

$$\sigma(A) = \{-4.9711,\ \ -0.0973,\ \ 2.0684\}.$$

Let the undesirable eigenvalues be $\{-0.0973,\ 2.0684\}$. For that, the Schur decomposition of matrix A is given by:

$$Q = \begin{bmatrix} 0.9851 & 0.1427 & 0.0962 \\ 0.0284 & -0.6864 & 0.7267 \\ -0.1697 & 0.7131 & 0.6802 \end{bmatrix},\quad Q_m = \begin{bmatrix} 0.1427 & 0.0962 \\ -0.6864 & 0.7267 \\ 0.7131 & 0.6802 \end{bmatrix}.$$

The corresponding matrices A_m and B_m are given by

$$A_m = \begin{bmatrix} -0.09726 & 0.1413 \\ 0 & 2.0684 \end{bmatrix}, \quad B_m = \begin{bmatrix} 0.713 & -0.6864 \\ 0.6802 & 0.7267 \end{bmatrix}.$$

The LMIs of Lemma 6.2 are not feasible. For this, choose Γ and 4 (η) diagonalizable and Hurwitz matrices H_i satisfying (6.56), (6.57), and (6.75) as follows:

$$H_1 = \begin{bmatrix} -4 & 1 \\ 0 & -5 \end{bmatrix}, \quad H_2 = \begin{bmatrix} -4.1 & 0.1 \\ 0 & -4.8 \end{bmatrix},$$
$$H_3 = \begin{bmatrix} -3.9 & 1 \\ 0 & -5.2 \end{bmatrix}, \quad H_4 = \begin{bmatrix} -4.2 & 0.1 \\ 0 & -5.1 \end{bmatrix},$$
$$\Gamma_1 = \begin{bmatrix} 4.1 & 1 \\ 2 & 6.3 \end{bmatrix}, \quad \Gamma_2 = \begin{bmatrix} 0.5 & 1.5 \\ 1 & 0.5 \end{bmatrix}.$$

The LMIs of Theorem 6.3 are feasible and yield the following P_m, F_m, and G_m:

$$P_m = \begin{bmatrix} 2.3587 & 0.1400 \\ 0.1400 & 15.5694 \end{bmatrix}, \; F_m = \begin{bmatrix} -0.2604 & -3.9212 \\ 0.1698 & -3.9833 \end{bmatrix},$$
$$G_m = \begin{bmatrix} -0.2608 & -2.0107 \\ 0.2148 & -2.0859 \end{bmatrix}.$$

From the Algorithm 6.1, the following matrices F, G, and P are obtained:

$$F = F_m Q_m^T = \begin{bmatrix} -0.4144 & -2.6708 & -2.8529 \\ -0.3590 & -3.0112 & -2.5883 \end{bmatrix},$$
$$G = G_m Q_m^T = \begin{bmatrix} -0.2307 & -1.2821 & -1.5537 \\ -0.1700 & -1.6632 & -1.2656 \end{bmatrix},$$
$$P = QRQ^T = \begin{bmatrix} 1.1664 & 0.8906 & 1.1148 \\ 0.8906 & 9.1946 & 6.5438 \\ 1.1148 & 6.5438 & 8.5676 \end{bmatrix},$$

where matrix P_o can be taken as any positive scalar, namely $P_o = 1$.

Figure 6.2 presents ellipsoid set of invariance and contractivity for the reduced-order saturated linear system computed with constraints on the rate, computed using Theorem 6.3, together with the set $\mathscr{L}(G_m)$ (in solid line) and $\mathscr{L}(F_m)$ (in dotted line) with 10 trajectories. The region of asymptotic stability where the saturation does not occur as a solution of the results of [12] is given by $\mathscr{L}(F_m)$ (in dotted line).

Figure 6.3 plots the evolution of the two components of the control rate for $z_m(0) = [-0.4470 \;\; 0.3860]^T$.

Example 6.2 In order to illustrate the use of the augmentation technique, consider the double integrator system in discrete-time:

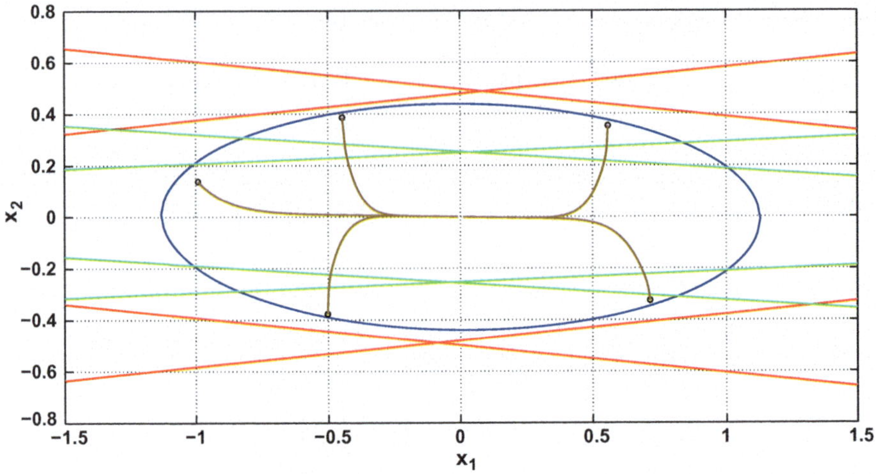

Fig. 6.2 Trajectories inside ellipsoid set of invariance together with the set $\mathscr{L}(G_m)$ (in *red*) and $\mathscr{L}(F_m)$ (in *green*)

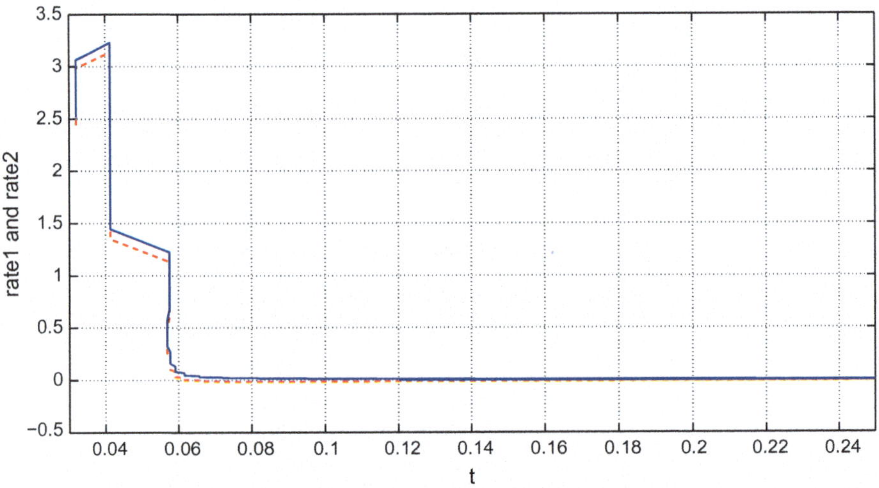

Fig. 6.3 Control rate evolution

$$A = \begin{bmatrix} 1 & 1 \\ 0 & 1 \end{bmatrix}, \quad B = \begin{bmatrix} 0.5 \\ 1 \end{bmatrix}, \quad \rho = 10.$$

Let the asymmetric constraints on the control increment be as follows:

$$q_1 = 15; \quad q_2 = 5.$$

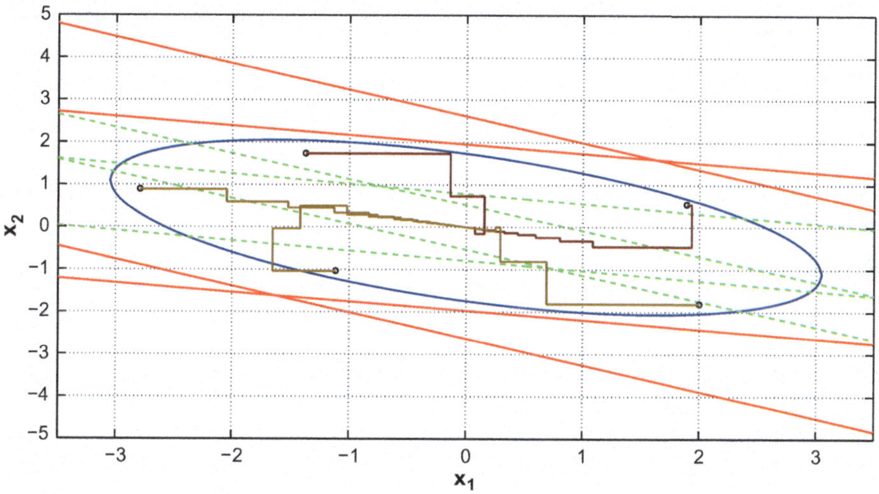

Fig. 6.4 Trajectories inside ellipsoid set of invariance together with the set $\mathscr{L}(G_a)$ (*solid line* in *red*) and $\mathscr{L}(K)$ (*dotted line* in *green*)

The open-loop eigenvalues of the system are given by:

$$\sigma(A) = \{1, \quad 1\}.$$

The open-loop system does not contain $n - m$ stable eigenvalues. To overcome this condition, the augmentation technique is used:

$$B_a = \begin{bmatrix} 0.5 & 0 \\ 1 & 0 \end{bmatrix}.$$

The LMIs of Lemma 6.2 are feasible but lead to a restrictive region of stability. Hence, choose Γ and 4 (η) matrices H_i satisfying (6.75) with given fictitious constraints on the increment $\varphi_1 = 10$ and $\varphi_2 = 6.5$ as follows:

$$H_1 = \begin{bmatrix} 0.45 & 0.05 \\ 0 & 0.5 \end{bmatrix}, \quad H_2 = \begin{bmatrix} 0.5 & 0.1 \\ 0 & 0.4 \end{bmatrix};$$
$$H_3 = \begin{bmatrix} 0.4 & 0.1 \\ 0 & 0.5 \end{bmatrix}, \quad H_4 = \begin{bmatrix} 0.6 & 0.1 \\ 0 & 0.4 \end{bmatrix},$$
$$\Gamma_1 = \begin{bmatrix} 1 & 1.2 \\ 2.5 & 1.75 \end{bmatrix}, \quad \Gamma_2 = \begin{bmatrix} 1.5 & 0.6 \\ 1.1 & 0.4 \end{bmatrix}.$$

Solving the LMIs of Theorem 6.3 yields the following solutions:

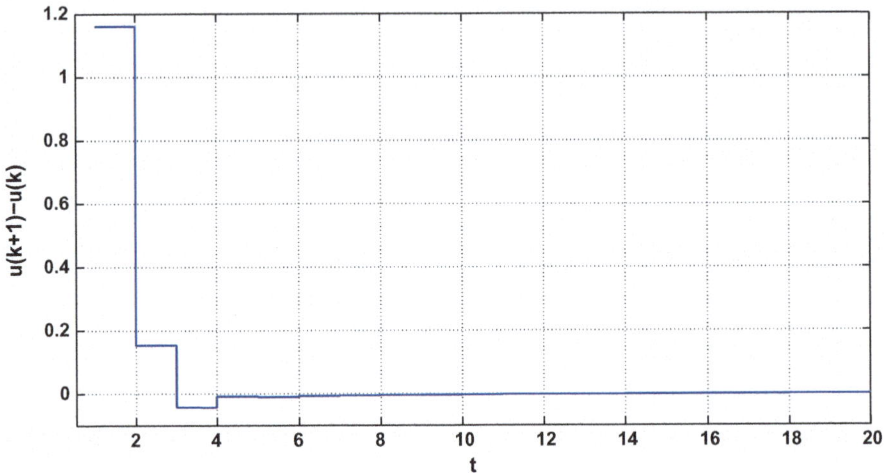

Fig. 6.5 Evolution of the increment of the control

$$K = \begin{bmatrix} -0.2993 & -1.2646 \\ 1.1513 & 1.8940 \end{bmatrix},$$

$$G_a = \begin{bmatrix} -0.1113 & -0.5092 \\ 0.2378 & 0.3810 \end{bmatrix},$$

$$P = \begin{bmatrix} 1.5072 & 1.1894 \\ 1.1894 & 3.3030 \end{bmatrix},$$

where G_a is the augmented matrix corresponding to matrix G. The effective gain matrices F and G are to be extracted from matrices K and G_a, respectively, as follows:

$$F = \begin{bmatrix} -0.2993 & -1.2646 \end{bmatrix}, G = \begin{bmatrix} -0.1113 & -0.5092 \end{bmatrix}.$$

One can note from Fig. 6.4 that the fictitious constraints φ_1 and φ_2, which were used as additional degrees of freedom to satisfy condition (6.31), limit in return the sets $\mathscr{L}(G)$ and $\mathscr{L}(F)$. Finally, the LMIs of Lemma 6.2 were also used leading to acceptable solutions with a reduced region of stability than the one obtained with the LMIs of Theorem 6.3. Figure 6.4 presents the evolution of the state of the system inside the ellipsoid set $\varepsilon(P, \rho)$. The region of asymptotic stability where the saturation does not occur as a solution of the results of [12] is given by $\mathscr{L}(F)$ (in dotted line). While Fig. 6.5 plots the evolution of the increment for initial condition $x_o = [-0.7650\ 1.8421]^T$.

6.5 Conclusion

In this chapter, a new design formulation of the stabilizing controller for linear system subject to actuator saturation and asymmetric constraints on its increment or rate is presented. This technique is based on the use of the reduced-order system and reduced dimension LMI to simplify the computations. The problem of asymmetric constraints on the control increment or rate is also taken into account by using the solution of the nonlinear algebraic equation (6.30) under LMI form. The obtained LMIs of Lemma 6.2, even with reduced dimensions, are relaxed considerably by the use of the idea of introducing admissible uncertainties in the control dynamic due to equalities (6.70). The results are given both for continuous-time and discrete-time systems. Finally, two illustrative examples that prove the improvements with the proposed approach are presented.

References

1. A. Benzaouia, C. Burgat, Regulator problem for linear discrete-time systems with non-symmetrical constrained control. Int. J. Control **48**(6), 2441–2451 (1988)
2. A. Benzaouia, S. El Faiz, The regulator problem for linear systems with constrained control: an LMI approach. IMA J. Math. Control Inf. **23**(3), 335–345 (2006)
3. A. Benzaouia, F. Tadeo, F. Mesquine, The regulator problem for linear systems with saturation on the control and its increments or rate: an LMI approach. IEEE Trans. Circuits Syst. I Fundam. Theory Appl. **53**(12), 2681–2691 (2006)
4. S. Boyd, L. EL Ghaoui, E. Feron, V. Balakrishnan, *Linear Matrix Inequalities in System and Control Theory* (SIAM, Philadelphia, PA, 1994)
5. Y.Y. Cao, Z. Lin, Stability analysis of discrete-time systems with actuator saturation by saturation dependent Lyapunov function, in *Proceedings of the 41th IEEE CDC* (Las Vegas, USA, 2002)
6. J.J Dongarra, S. Hammarling, J. Wilkinson, Numerical considerations in computing invariant subspaces. SIAM J. Matrix Anal. Appl. **13**(1), 145–161 (1992)
7. G. Feng, M. Palaniswami, Y. Zhu, Stability of rate constrained robust pole placement adaptive control systems. Syst. Control Lett. **18**, 99–107 (1992)
8. T. Hu, Z. Lin, The equivalence of several set invariance conditions under saturation, in *Proceedings of the 41th IEEE CDC* (Las Vegas, USA, 2002)
9. T. Hu, Z. Lin, B.M. Chen, Analysis and design for discrete-time linear systems subject to actuator saturation. Syst. Control Lett. **45**, 97–112 (2002)
10. T. Hu, Z. Lin, B.M. Chen, An analysis and design method for linear systems subject to actuator saturation and disturbance. Automatica **38**, 351–359 (2002)
11. P. Kapasouris, M. Athans, Control systems with rate and magnitude saturation for neutrally stable open loop systems, in *Proceedings of the IEEE CDC* (Honolulu, Hawaii, U.S.A., 1990), pp. 3404–3409
12. F. Mesquine, F. Tadeo, A. Benzaouia, Regulator problem for linear systems with constraints on the control and its increments or rate. Automatica **40**(8), 1378–1395 (2004)
13. F. Mesquine, F. Tadeo, A. Benzaouia, Regulator constrained control and rate problem for linear systems with additive disturbances, in *American Conference on Control* (Boston, 2004), June 30–July 2
14. L. Trygve, R. Murray, T.Y. Fossen, Stabilization of integrator chains in the presence of magnitude and rate saturation; a gain scheduling approach, in *Proceedings of the IEEE CDC* (San Diego, USA, 1997), pp. 4004–4005

Chapter 7
Output Feedback Stabilization for Constrained Control Systems

7.1 Introduction

The problem of stabilizing linear systems by output feedback, despite its apparent simplicity, is still open. A number of numerical procedures have been proposed for solving the problem since the work of Kimura [1]. A survey was given by [2], and recent progress has been made for the related problem of pole placement; see [3–5] and the references therein. However, less works were proposed for linear systems with actuator saturation. In [6] a dynamic output feedback is considered while in [7, 8] and [9], the positive invariance approach is used. Static output feedback problem for both discrete-time and continuous-time linear systems subject to actuator saturation is studied by extending the results of [3–6, 8, 10–14] where a state feedback is used. The synthesis of the controller by static output feedback is also proposed by means of LMI's for discrete-time and continuous-time linear systems subject to actuator saturation. The obtained region of invariance and contractivity are generally less conservative.

It is true that the use of output feedback was neglected at the benefit of the observers when the state is not available for measure. However, in general case, the output is measured and hence easily available for feedback. State feedback and observer are the essential parts of this book but we aim to present also the case of output feedback that can be very useful in some cases, especially when the output matrix C is of full rank. Other results have been obtained by output feedback synthesis for sampled-data system with input saturation [12].

7.2 Problem Formulation

Consider the class of linear systems described by:

$$\delta x(\cdot) = Ax(\cdot) + Bsat(u(\cdot)) \tag{7.1}$$

$$y(\cdot) = Cx(\cdot) \tag{7.2}$$

A. Benzaouia et al., *Saturated Control of Linear Systems*, Studies in Systems, Decision and Control 124, https://doi.org/10.1007/978-3-319-65990-9_7

where $x(\cdot) \in \mathbb{R}^n$ is the state, $u(\cdot) \in \mathbb{R}^m$ is the control with $m \leq n$, $y(\cdot) \in \mathbb{R}^{p_o}$ is the output vector. The δ operator is defined as:

$$\delta x(\cdot) = \begin{cases} \dot{x}(t) & \text{for the continuous-time case} \\ x(k+1) & \text{for the discrete-time case} \end{cases} \tag{7.3}$$

Further, $sat(\cdot)$ is the standard saturation function assumed here to be normalized, *i. e.*, $|sat(u)| = min\{1, |u|\}$.
Let the following habitual assumptions for the problem hold:

- $AS1$: matrix C has full rank;
- $AS2$: (A, B) is stabilizable; (C, A) is detectable;

Consider the following static output feedback control law:

$$u(\cdot) = Ky(\cdot) = Fx(\cdot), F = KC \tag{7.4}$$

The closed-loop system is then given by,

$$\delta x(\cdot) = Ax(\cdot) + Bsat(KCx(\cdot)) \tag{7.5}$$

In this work, we are interested to the synthesis of a stabilizing controller for this class of linear systems subject to actuator saturation in both discrete-time and continuous-time cases. To this end let us recall the sets $\varepsilon(P, \rho)$ and $\mathscr{L}(H)$ as follows:

$$\varepsilon(P, \rho) = \left\{x \in \mathbb{R}^n \;/ x^T Px \leq \rho\right\} \tag{7.6}$$

$$\mathscr{L}(H) = \left\{x \in \mathbb{R}^n \;\; /|H_i x| \leq 1\right\}, \tag{7.7}$$

H_i the ith row of matrix H.

7.3 Output Feedback for Saturated Discrete-Time Linear Systems

The design of the stabilizing controller for the class of linear systems with actuator saturation is presented by using both the results of Theorem 1.10 and Lemma 1.3. A different proof of Theorem 1.10 is also proposed.

Theorem 7.1 *For a given positive scalar ρ, if there exist a symmetric matrix P and a matrix H such that,*

$$\begin{bmatrix} P & [A + B(D_i KC + D_i^- H)]^T P \\ * & P \end{bmatrix} \succ 0, \tag{7.8}$$

and,

$$\varepsilon(P,\rho) \subset \mathscr{L}(H), \forall i \in \mathscr{I} \tag{7.9}$$

then, the closed-loop system (7.5) is asymptotically stable $\forall x_o \in \varepsilon(P,\rho)$, *with* $\varepsilon(P,\rho)$ *and* $\mathscr{L}(H)$ *are defined by (7.6) and (7.7) respectively.*

Proof Assume that there exists a matrix H and a symmetric matrix P such that conditions (7.8) and (7.9) hold. Using Lemma 1.3, one can rewrite the saturated system (7.5) as follows: there exist $\gamma_1, \ldots, \gamma_\eta$ such that,

$$sat(KCx_k) = \sum_{i=1}^{\eta} \gamma_i(k)[D_i KC + D_i^- H]x_k;$$

$$\gamma_i(k) \geq 0, \sum_{i=1}^{\eta} \gamma_i(k) = 1, \tag{7.10}$$

The closed-loop system can be rewritten as,

$$x_{k+1} = \sum_{i=1}^{\eta} \gamma_i(k) A_i^{cl} x_k; \tag{7.11}$$

$$A_i^{cl} = A + B(D_i KC + D_i^- H), i \in \mathscr{I}$$

Consider now the Lyapunov function candidate given by:

$$V(x_k) = x_k^T P x_k \tag{7.12}$$

Its rate of increase on the trajectories of the system (7.11) is given by,

$$\begin{aligned} \Delta V(x_k) &= x_{k+1}^T P x_{k+1} - x_k^T P x_k \\ &= x_k^T \left\{ \left[\sum_{i=1}^{\eta} \gamma_i(k) A_i^{cl} \right]^T P \left[\sum_{i=1}^{\eta} \gamma_i(k) A_i^{cl} \right] - P \right\} x_k \end{aligned} \tag{7.13}$$

Let condition (7.8) be satisfied. Pre-multiply each inequality (7.8) for $i = 1, \ldots, \eta$ by $\gamma_i(k)$ and sum up the obtained inequalities. Bearing in mind that, $\sum_{i=1}^{\eta} \gamma_i(k) = 1$, one gets:

$$\begin{bmatrix} P & \left[\sum_{i=1}^{\eta} \gamma_i(k) A_i^{cl}\right]^T P \\ * & P \end{bmatrix} \succ 0, \tag{7.14}$$

The use of Schur complement allows one to write condition (7.14) under the equivalent form,

$$\left[\sum_{i=1}^{\eta} \gamma_i(k) A_i^{cl}\right]^T P \left[\sum_{i=1}^{\eta} \gamma_i(k) A_i^{cl}\right] - P \prec 0, \quad (7.15)$$

which ensures that,

$$\Delta V(x_k) < -\gamma(\| x_k \|); \text{ where}$$

$$\gamma(\| x_k \|) = \min_i \lambda_{min}(P - \left[\sum_{i=1}^{\eta} \gamma_i(k) A_i^{cl}\right]^T P \left[\sum_{i=1}^{\eta} \gamma_i(k) A_i^{cl}\right]) \| x_k \|^2 \quad (7.16)$$

Taking into account condition (7.9) and noticing that $\varepsilon(P, \rho)$ is a contractively invariant set, one can guarantee that for all $x_o \in \varepsilon(P, \rho) \subset \mathscr{L}(H)$, the saturated system (7.5) is asymptotically stable. □

Note that Theorem 7.1 proposes a different proof of Theorem 1.10 of [14] for linear systems with state feedback by letting $C = \mathbb{I}$. This result of stability can be exploited in the synthesis of the controller by the following result.

Theorem 7.2 *For a given positive scalar ρ, if there exist symmetric positive definite matrix $X \in \mathbb{R}^{n \times n}$, matrices $V \in \mathbb{R}^{p_o \times p_o}$, $Y \in \mathbb{R}^{m \times p_o}$ and $Z \in \mathbb{R}^{m \times n}$ solutions of the following LMI's:*

$$\begin{bmatrix} X & (AX + BD_i YC + BD_i^- Z)^T \\ * & X \end{bmatrix} \succ 0, \quad (7.17)$$

$$\begin{bmatrix} \frac{1}{\rho} & Z_j \\ * & X \end{bmatrix} \succ 0, \quad (7.18)$$

$$VC - CX = 0, \quad (7.19)$$

$$\forall i \in \mathscr{I}, \forall j \in [1, m]$$

where Z_j is the jth row of matrix Z; then the closed-loop system subject to saturation (7.5) is asymptotically stable at the origin $\forall x_o \in \varepsilon(P, \rho)$ with,

$$K = YV^{-1} \quad (7.20)$$

$$H = ZX^{-1} \quad (7.21)$$

$$P = X^{-1} \quad (7.22)$$

Proof For all $i \in \mathscr{I}$, the LMI (7.8) can be transformed by Schur complement to the following:

$$P\left[A + B(D_i KC + D_i^- H)\right] P^{-1} \left[A + B(D_i KC + D_i^- H)\right]^T P - P \prec 0 \quad (7.23)$$

Post-multiplying and pre-multiplying the last inequalities by P^{-1}, leads to:

$$\left[A + B(D_i KC + D_i^- H)\right] P^{-1} \left[A + B(D_i KC + D_i^- H)\right]^T - P^{-1} \prec 0 \quad (7.24)$$

letting $X = P^{-1}$, inequality (7.24) becomes,

$$\left[A + B(D_i KC + D_i^- H)\right] X \left[A + B(D_i KC + D_i^- H)\right]^T - X \prec 0$$

The use of the Schur complement a second time leads to:

$$\begin{bmatrix} X & (AX + BD_i KCX + BD_i^- HX)^T \\ * & X \end{bmatrix} \succ 0 \quad (7.25)$$

According to Eq. (7.19), one can write $KCX = KVC$. By virtue of assumptions $(AS1 - AS2)$, and as matrix X is positive definite, matrix V solution of (7.19) is non-singular. Further, by letting $KV = Y$ and $HX = Z$, the LMI (7.17) follows together with relations (7.20) and (7.21). Using [13], the inclusion condition (7.9) can also be transformed to the equivalent LMI (7.18) with $X = P^{-1}$. □

It is worth noting that the state feedback problem follows readily from Theorem 7.2 by letting $C = \mathbb{I}_m$. In this case, $V = X$. The resolution of these LMI's can be extended to the case where the scalar ρ is also taken as a design variable.

The conditions of Theorem 7.2 are in fact more conservative due to Eq. (7.19). In order to relax this conservatism, we associate this equation to a second matrix S, as a slack variable, different from X as suggested by [15] and [10] for unsaturated systems. This technique is presented by the following result.

Theorem 7.3 *For a given scalar ρ, if there exist symmetric matrix $X \in \mathbb{R}^{n \times n}$, matrices $V \in \mathbb{R}^{p_o \times p_o}$, $S \in \mathbb{R}^{n \times n}$, $Y \in \mathbb{R}^{m \times p_o}$ and $Z \in \mathbb{R}^{m \times n}$ solutions of the following LMI's:*

$$\begin{bmatrix} S + S^T - X & \left[AS + BD_i YC + BD_i^- Z\right]^T \\ * & X \end{bmatrix} \succ 0, \quad (7.26)$$

$$\begin{bmatrix} \frac{1}{\rho} & Z_j \\ * & S + S^T - X \end{bmatrix} \succ 0, \quad (7.27)$$

$$VC - CS = 0, \quad (7.28)$$

$$\forall i \in \mathscr{I}, \forall j \in [1, m]$$

where Z_j is the jth row of matrix Z; then the closed-loop system subject to saturation (7.5) is asymptotically stable at the origin $\forall x_o \in \varepsilon(P, \rho)$, with,

$$K = YV^{-1} \quad (7.29)$$

$$H = ZS^{-1} \quad (7.30)$$

$$P = X^{-1} \quad (7.31)$$

Proof The main idea of this proof is given by [10]. Using Eqs. (7.28), (7.29), and (7.30), the LMI (7.26) can be rewritten equivalently as,

$$\begin{bmatrix} S + S^T - X & S^T(A + BD_i KC + BD_i^- H)^T \\ * & X \end{bmatrix} \succ 0,$$

It is obvious that if (7.26) holds, then, $S + S^T - X \succ 0$, thus, matrix S is non-singular and V is also non-singular. Since matrix X is positive definite, we have,

$$(X - S)^T X^{-1}(X - S) \succ 0, \tag{7.32}$$

this implies that,

$$S^T X^{-1} S \succ S + S^T - X, \tag{7.33}$$

Inequality (7.33) with (7.26) enable us to write,

$$\begin{bmatrix} S^T X^{-1} S & S^T(A + BD_i KC + BD_i^- H)^T \\ * & X \end{bmatrix} \succ 0$$

This LMI is equivalent to:

$$\begin{bmatrix} S^T & 0 \\ 0 & X \end{bmatrix} \begin{bmatrix} X^{-1} & (A + BD_i KC + BD_i^- H)^T X^{-1} \\ * & X^{-1} \end{bmatrix} \begin{bmatrix} S & 0 \\ 0 & X \end{bmatrix} \succ 0$$

To complete the proof, we show thereafter that the inclusion (7.9) is equivalent to

$$H_j P^{-1} H_j^T \leq \frac{1}{\rho}$$

using (7.30) and (7.31), one can write

$$(ZS^{-1})_j X (ZS^{-1})_j^T < \frac{1}{\rho}$$

by Schur complement, condition above becomes:

$$\begin{bmatrix} \frac{1}{\rho} & (ZS^{-1})_j \\ * & X^{-1} \end{bmatrix} \succ 0$$

bearing in mind that $(ZS)_j = Z_j S$, one obtains:

$$\begin{bmatrix} \frac{1}{\rho} & Z_j S^{-1} \\ * & X^{-1} \end{bmatrix} \succ 0,$$

pre-multiplying by $diag\{\mathbb{I}, S^T\}$ and post-multiplying $diag\{\mathbb{I}, S\}$ the obtained inequality, leads to

$$\begin{bmatrix} \frac{1}{\rho} & Z_j \\ * & S^T X^{-1} S \end{bmatrix} \succ 0$$

following the same reasoning for developing $S^T X^{-1} S$, LMI (7.27) is obtained. By letting $P = X^{-1}$, the LMI (7.26) is the sufficient condition of stability of Theorem 7.1 which, together with condition (7.27), ensure that the closed-loop system is asymptotically stable at the origin $\forall x_o \in \varepsilon(P, \rho)$. □

As in the previous works on the problem of saturated systems, it is of great interest to obtain the largest ellipsoid $\varepsilon(P, \rho)$ of initial conditions. It is worth noting that in the case of output feedback we have more constraining equalities (7.19). We present hereafter two optimization procedures to obtain such ellipsoid by writing $\mu = 1/\rho$:

$$(Pb.1) : \begin{cases} \min_{(X,Y,V,Z)}(\mu) \\ s.t. \ (7.17), (7.27)(7.19). \end{cases}$$

This optimization problem, when is feasible, can help to enlarge the ellipsoids $\varepsilon(P, \rho)$ by maximizing the scalar ρ. Since the volume of the ellipsoid is proportional to the trace of matrix X, a second way to obtain larger sets of invariance and contractivity is to solve the following optimization problem:

$$(Pb.2) : \begin{cases} \sup_{(X,Y,V,Z)} Trace(X) \\ s.t. \ (7.17), (7.18), (7.19) \end{cases}$$

When this optimization problem is feasible, the volume of the obtained ellipsoids is maximum with respect to the data of the system.

Example 7.1 Consider the double integrator system modeled by the saturated discrete-time linear system given by the following matrices:

$$A = \begin{bmatrix} 1 & 1 \\ 0 & 1 \end{bmatrix}; B = \begin{bmatrix} 0.5 \\ 1 \end{bmatrix}; C = \begin{bmatrix} 1 & 1 \end{bmatrix}$$

For this example with $n = 2$, and $m = p_o = 1$, we have to solve 7 LMI's with 4 variables. Let the scalar ρ be given equal to 1. The use of the LMI's Matlab Toolbox leads to the following results.

$$P = \begin{bmatrix} 0.3661 & 0.1494 \\ 0.1494 & 0.3661 \end{bmatrix}; \ H = \begin{bmatrix} -0.3402 & -0.5756 \end{bmatrix}; \ K = -0.7826;$$

$$A_1^{cl} = A + BH = \begin{bmatrix} 0.8299 & 0.7122 \\ -0.3402 & 0.4244 \end{bmatrix};$$

$$A_2^{cl} = A + BKC = \begin{bmatrix} 0.6087 & 0.6087 \\ -0.7826 & 0.2174 \end{bmatrix};$$

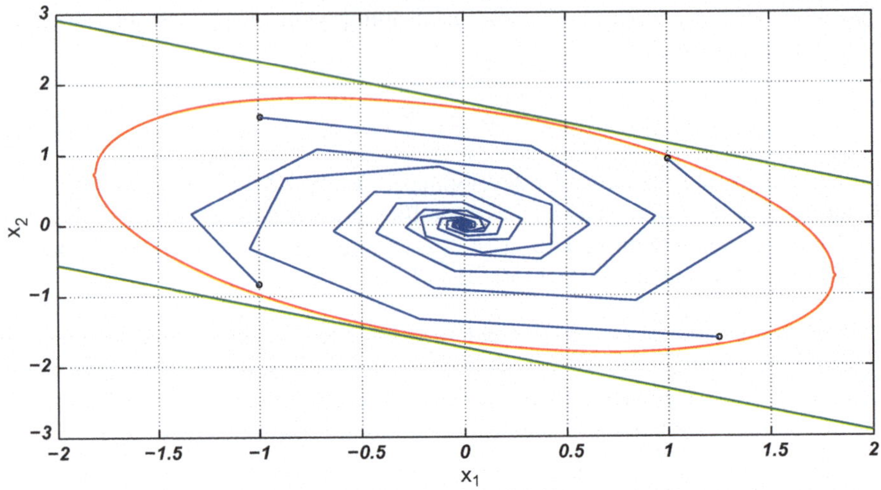

Fig. 7.1 4 Trajectories inside the invariance and contractivity ellipsoid for the system by output feedback

$$\sigma(A_1^{cl}) = \{0.6271 + 0.4485i; 0.6271 - 0.4485i\};$$
$$\sigma(A_2^{cl}) = \{0.4131 + 0.6619i; 0.4131 - 0.6619i\}.$$

Figure 7.1 presents the ellipsoid set of invariance and contractivity for the saturated discrete-time linear system with output feedback.

Example 7.2 Consider now the following example studied by [15] where $C = \mathbb{I}_3$ but here we consider different matrix C:

$$A = \begin{bmatrix} 1 & 1 & 0.5 \\ 0 & 1 & 1 \\ 1 & 0 & 1 \end{bmatrix}; B = \begin{bmatrix} 1.67 \\ 0.5 \\ 1 \end{bmatrix}; C = \begin{bmatrix} 1 & 1 & 0 \\ 0 & 0 & 1 \end{bmatrix}$$

For this example with $n = 3$, $m = 1$ and $p_o = 2$, Let the scalar ρ be given equal to 1. First, we try to solve the LMI's of Theorem 7.2. Unfortunately they are not feasible. However, and as it is expected while using the LMI's of Theorem 7.3 the problem becomes feasible. In fact it leads to the following results given by:

$$P = \begin{bmatrix} 0.1973 & 0.1697 & 0.2110 \\ 0.1697 & 0.1693 & 0.1970 \\ 0.2110 & 0.1970 & 0.2491 \end{bmatrix};$$
$$H = \begin{bmatrix} -0.3691 & -0.3310 & -0.4598 \end{bmatrix};$$
$$K = [-0.5299 - 0.6709];$$

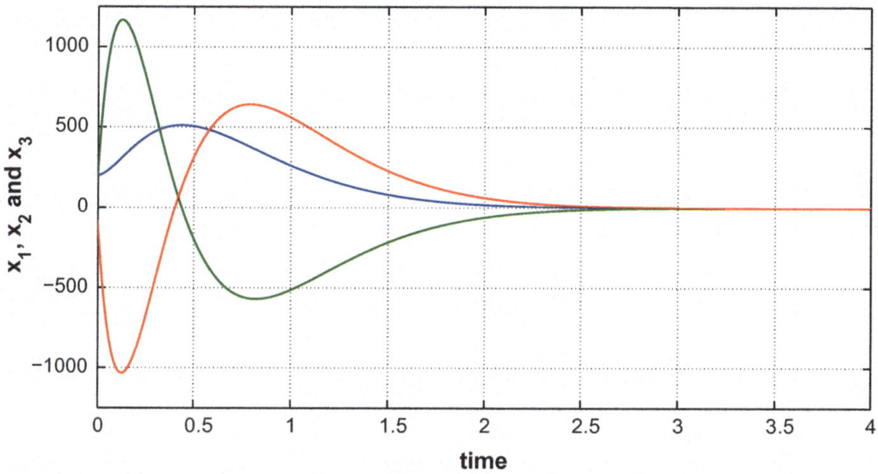

Fig. 7.2 Trajectories of the saturated system with output feedback obtained with Theorem 7.3

$$A_1^{cl} = A + BH = \begin{bmatrix} 0.3836 & 0.4472 & -0.2678 \\ -0.1845 & 0.8345 & 0.7701 \\ 0.6309 & -0.3310 & 0.5402 \end{bmatrix};$$

$$A_2^{cl} = A + BKC = \begin{bmatrix} 0.1150 & 0.1150 & -0.6205 \\ -0.2650 & 0.7350 & 0.6645 \\ 0.4701 & -0.5299 & 0.3291 \end{bmatrix};$$

$$\sigma(A_1^{cl}) = \{0.4186 + 0.7336i; 0.4186 - 0.7336i; 0.9212\};$$

$$\sigma(A_2^{cl}) = \{0.3152; 0.4319 + 0.7614i; 0.4319 - 0.7614i\}$$

$$VC - CS = 10^{-13} * \begin{bmatrix} 0 & -0.1421 & 0.4263 \\ 0.0711 & -0.2132 & 0 \end{bmatrix}$$

Figures 7.2 and 7.3 present, respectively, the evolution of state and control components. It is clear that the case of Theorem 7.3 is more favorable since it gives a larger ellipsoid.

7.4 Output Feedback for Saturated Continuous-Time Linear Systems

In what follows, results obtained for discrete-time systems are extended to the case of continuous-time systems. It is a straightforward extension, but it is presented here for completeness. Hence, stabilizing controller design for the class of continuous-time linear systems with actuator saturation by static output feedback law is presented.

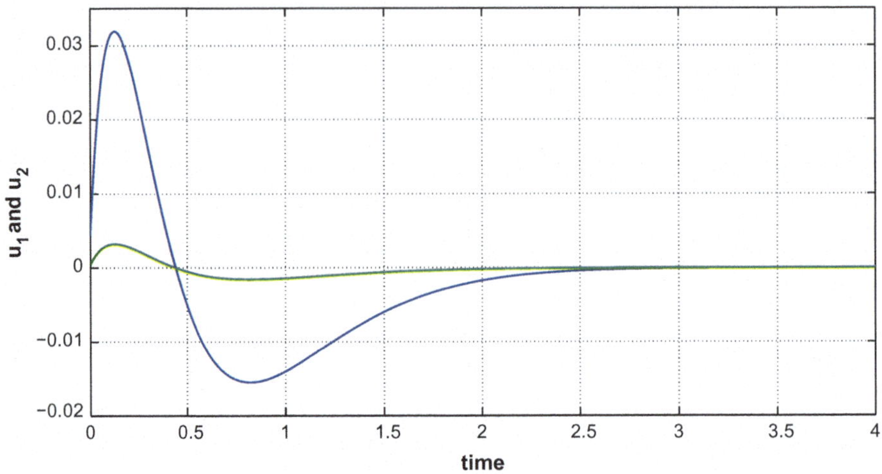

Fig. 7.3 Control components using Theorem 7.3

Consider the class of linear continuous-time systems described above and given by (7.2) satisfying assumptions $AS1$ and $AS2$ as:

$$\dot{x}_(t) = Ax(t) + Bsat(u(t)) \quad (7.34)$$

$$y(t) = Cx(t) \quad (7.35)$$

where $x(t) \in \mathbb{R}^n$ is the state, $u(t) \in \mathbb{R}^m$ is the control with $m \leq n$, $y(t) \in \mathbb{R}^{p_o}$ is the output vector.
With a static output feedback control law, one obtains:

$$u(t) = Ky(t) = Fx(t), \; F = KC \quad (7.36)$$

The closed-loop system is then given by,

$$\dot{x}(t) = Ax(t) + Bsat(KCx(t)) \quad (7.37)$$

Theorem 7.4 *For a given positive scalar ρ, if there exist a symmetric positive definite matrix P and a matrix H such that,*

$$[A + B(D_s KC + D_s^- H)]^T P + P[A + B(D_s KC + D_s^- H)] \prec 0, \quad (7.38)$$

and,

$$\varepsilon(P, \rho) \subset \mathscr{L}(H), \forall s \in \mathscr{I} \quad (7.39)$$

then, the closed-loop system (7.37) is asymptotically stable $\forall x_o \in \varepsilon(P, \rho)$, *with* $\varepsilon(P, \rho)$ *and* $\mathscr{L}(H)$ *are defined by (7.6) and (7.7) respectively.*

Proof Assume that there exists a matrix H and a symmetric matrix P such that conditions (7.38) and (7.39) hold. Using Lemma 1.3, one can rewrite the saturated system (7.37) as follows: there exist $\gamma_1, \ldots, \gamma_\eta$ such that,

$$sat(KCx(t)) = \sum_{s=1}^{\eta} \gamma_s(t)[D_s KC + D_s^- H]x(t); \tag{7.40}$$

$$\gamma_s(t) \geq 0, \sum_{s=1}^{\eta} \gamma_s(t) = 1, \eta = 2^m.$$

The closed-loop system can be rewritten as,

$$\begin{aligned} \dot{x}(t) &= \sum_{s=1}^{\eta} \gamma_s(t) A_s^{cl} x(t); \\ A_s^{cl} &= A + B(D_s KC + D_s^- H), s \in \mathscr{I} \end{aligned} \tag{7.41}$$

Consider now the Lyapunov function candidate given by:

$$V(x(t)) = x(t)^T P x(t) \tag{7.42}$$

The derivative of Lyapunov function on the trajectories of the system (7.41) is given by:

$$\begin{aligned} \dot{V}(x(t)) &= \dot{x}(t)^T P x(t) + x(t)^T P \dot{x}(t) \\ &= x(t)^T \left[\sum_{s=1}^{\eta} \gamma_s(t) A_s^{cl} \right]^T P x(t) + x(t)^T P \left[\sum_{s=1}^{\eta} \gamma_i(t) A_s^{cl} \right] x(t) \end{aligned}$$

Let condition (7.38) be satisfied. Pre-multiply each inequality (7.38) for $s = 1, \ldots, \eta$ by $\gamma_s(t)$ and sum up the obtained inequalities. Bearing in mind that, $\sum_{s=1}^{\eta} \gamma_s(t) = 1$, one gets:

$$\left[\sum_{s=1}^{\eta} \gamma_s(t) A_s^{cl} \right]^T P + P \left[\sum_{s=1}^{\eta} \gamma_s(t) A_s^{cl} \right] \prec 0, \tag{7.43}$$

where:

$$A_s^{cl} = A + B(D_s KC + D_s^- H) \tag{7.44}$$

$$\sum_{s=1}^{\eta} \gamma_s(t)[(A_s^{cl})^T P + P A_s^{cl}] \prec 0, \forall s \in \mathcal{I}$$

which ensures that,

$$\dot{V}(x(t)) < -\gamma(\| x(t) \|);$$

where

$$\gamma(\| x(t) \|) = \min_i \lambda_{min}(\left[\sum_{s=1}^{\eta} \gamma_s(t) A_s^{cl}\right]^T P + P\left[\sum_{s=1}^{\eta} \gamma_s(t) A_s^{cl}\right]) \| x(t) \|^2 \quad (7.45)$$

Taking into account condition (7.9) and noticing that $\varepsilon(P, \rho)$ is a contractively invariant set, one can guarantee that for all $x_o \in \varepsilon(P, \rho) \subset \mathcal{L}(H)$, the saturated system (7.5) is asymptotically stable. □

Theorem 7.4 can not be used as it is for synthesis, to do so we propose the following:

Corollary 7.1 *For a given positive scalar ρ, if there exist a symmetric positive definite matrix X and matrices V, Y and Z such that $\forall s \in \mathcal{I}$,*

$$\left[AX + BD_sYC + BD_s^- Z\right]^{sym} \prec 0 \quad (7.46)$$

$$VC = CX \quad (7.47)$$

$$\begin{bmatrix} 1/\rho & Z_j \\ * & X \end{bmatrix} \succ 0, \quad (7.48)$$

then, the closed-loop system (7.37) is asymptotically stable $\forall x_o \in \varepsilon(P, \rho)$, with $\varepsilon(P, \rho)$ and $\mathcal{L}(H)$ defined by (7.6) and (7.7) respectively. Further, the stabilizing output feedback is given by

$$K = YV^{-1}, \quad H = ZX^{-1} \quad (7.49)$$

Proof Assume conditions 7.4, (7.47) and (7.48) hold true. Noting that $P = X^{-1}$. Replace Y and Z by KV and HX respectively. Use $VC = CX$ and multiply by P at the right and the left. One obtains the condition (7.38) of Theorem 7.1. (7.48) induces the inclusion (7.39) needed to complete the proof. □

Example 7.3 Consider now the following example:

$$A = \begin{bmatrix} -0.5 & 0.5 & 0.2 \\ 0 & 0.2 & 1 \\ 0 & 0 & -0.3 \end{bmatrix}; B = \begin{bmatrix} 1.67 \\ 0.5 \\ 1 \end{bmatrix}; C = \begin{bmatrix} 1 & 1 & 0 \\ 0 & 0 & 1 \end{bmatrix}$$

For this example with $n = 3$, $m = 1$ and $p_o = 2$, we have to solve the LMI's of Theorem 7.4. Let the scalar ρ be given equal to 10. The use of the LMI's Matlab

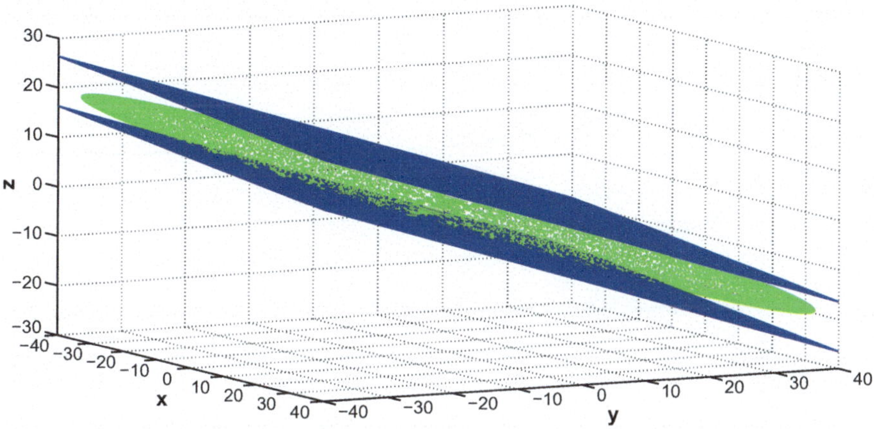

Fig. 7.4 Domains $\varepsilon(P, \rho)$ in *green* and $\mathcal{L}(H)$ in *blue*

Toolbox leads to the following results:

$$P = \begin{bmatrix} 0.0920 & -0.0071 & 0.1651 \\ -0.0071 & 0.0920 & 0.1651 \\ 0.1651 & 0.1651 & 0.6702 \end{bmatrix};$$
$$K = \begin{bmatrix} -2.5065 & -9.7307 \end{bmatrix};$$
$$H = \begin{bmatrix} -0.0187 & -0.0872 & -0.1979 \end{bmatrix}$$

Figures 7.4, 7.5 and 7.6 present respectively the domains for initial conditions, time evolution for states components and control evolution.

Example 7.4 Consider the following second example studied in [16] and given by:

$$\dot{x}(t) = Ax(t) + Bsat(u(t)), \tag{7.50}$$

with

$$A = \begin{bmatrix} -0.6161 & 0.5194 & 0.1941 & -0.0348 & -0.0517 & -0.0208 & 0 \\ 2.0000 & 0 & 0 & 0 & 0 & 0 & 0 \\ 0 & 2.0000 & 0 & 0 & 0 & 0 & 0 \\ 0 & 0 & 1.0000 & 0 & 0 & 0 & 0 \\ 0 & 0 & 0 & 1.0000 & 0 & 0 & 0 \\ 0 & 0 & 0 & 0 & 0.5000 & 0 & 0 \\ 0 & 0 & 0 & 0 & 0 & 0.1250 & 0 \end{bmatrix}; B = \begin{bmatrix} 1 \\ 0 \\ 0 \\ 0 \\ 0 \\ 0 \\ 0 \end{bmatrix};$$
$$C = \begin{bmatrix} -0.0966 & -0.7981 & -0.4005 & 0.0320 & 0.1098 & 0.1247 & 0.1642 \end{bmatrix}$$

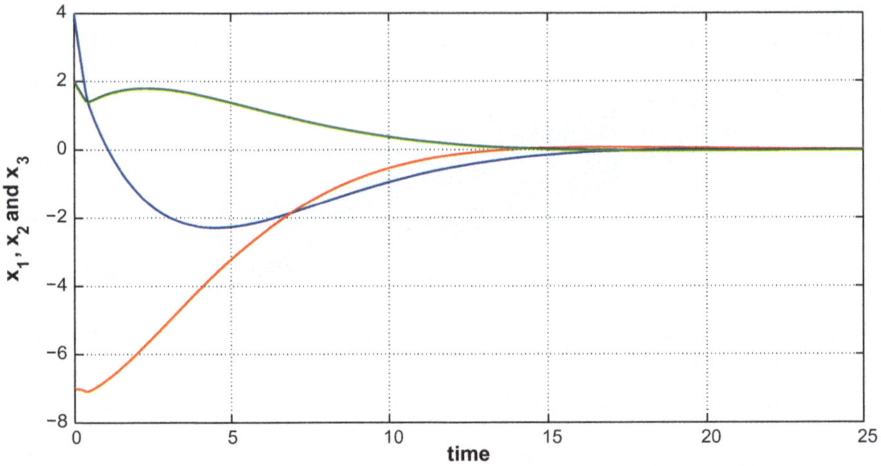

Fig. 7.5 States evolution

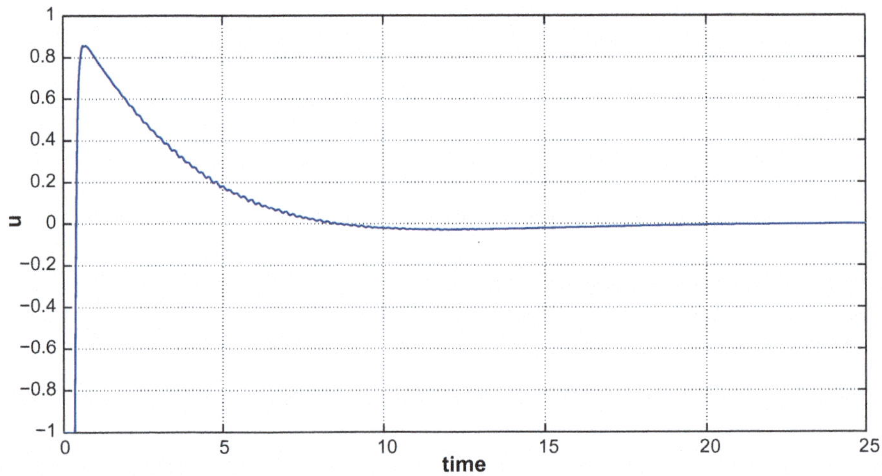

Fig. 7.6 Control evolution

Since the proposed method works only with symmetric constraints on the control, let the control be constrained between -1 and 1. Note that another approach which is more suitable with non-symmetrical constraints, the so-called invariance positive approach was also tried, but unfortunately, it was not successful due to the particular form of matrix B. The solution of the proposed LMI's of Theorem 7.4 computed with Scilab leads to the following results.

$$P = 10^{-3} \times \mathbb{I}_7; \; K = 8.4 \times 10^{-5}; \; \rho = 100.$$

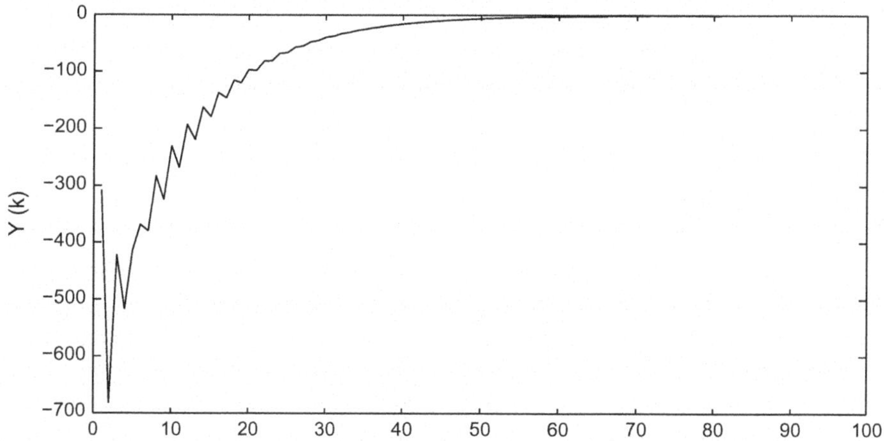

Fig. 7.7 Output evolution

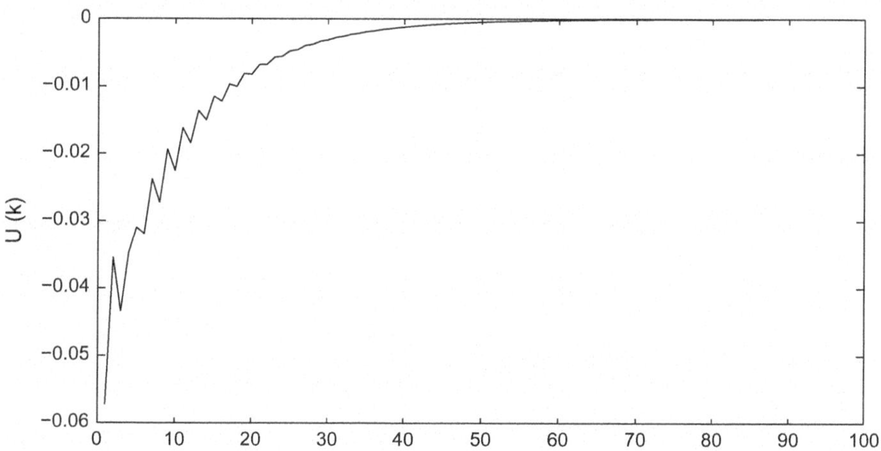

Fig. 7.8 Input evolution

The initial value must be chosen such that, $x_o^T P x_o \leq \rho$. Consider the following initial value which respects the condition choice, that means that the initial value belongs to the ellipsoid set of asymptotic stability and where the saturation is taken into account.

$$x_o = [300, 300, 0, 300, 0, 0, -300]^T$$

The evolution of the output of the process is plotted in Fig. 7.7 while the evolution of the control is plotted in Fig. 7.8. Note that this control satisfies the severe constraints even this small constraint was not taken into account in the design method of the controller. It was possible only by the choice of the initial state inside the ellipsoid set which was built according to a symmetric constraint equal to 1.

7.5 Conclusion

This chapter deals with sufficient conditions of asymptotic stability of static output feedback for both cases of discrete-time and continuous-time linear systems subject to actuator saturation. The results of [14] obtained with state feedback control are extended to the case of output feedback control. A simple proof, in the discrete-time case, based on manipulations of LMI's with Schur complement, together with the idea of Lemma 1.3, is also used. Two examples are studied to illustrate this approach. Further, extension to continuous-time case is easily obtained and proposed to complete the chapter. An example is also presented in this case.

Other techniques using dead-zone loops were developed for saturated linear systems by output feedback [11].

References

1. H. Kimura, Pole assignment by gain output feedback. IEEE Trans. Aut. Control **20**(4), 509–516 (1975)
2. V.L. Syrmos, C. Abdallah, P. Dorato, Static output feedback: a survey, in *Proceedings of IEEE CDC, Florida, December 14–16* (1994), pp. 837–842
3. C.A.R. Crusius, A. Trofino, Sufficient LMI conditions for output feedback control problems. IEEE Trans. Aut. Control **44**(5), 1053–1057 (1999)
4. L. El Ghaoui, F. Oustry, M. Ait Rami, Cone complementarity algorithm for static output-feedback and related problems. IEEE Trans. Control **42**(8), 1171–1176 (1997)
5. G. Garcia, P. Pradin, F. Zeng, Stabilization of discrete-time linear systems by static output feedback. IEEE Trans. Aut. Control **46**(12), 1954–1958 (2001)
6. D. Henrion, S. Tarbouriech, G. Garcia, Output feedback robust stabilization of uncertain linear systems with saturating controls: an LMI approach. IEEE Trans. Aut. Control **44**(11), 509–516 (1999)
7. A. Benzaouia, D. Mehdi, The output feedback saturated controller design for linear systems, in *MED Conference, Portugal* (2002)
8. E.B. Castelan, S. Tarbouriech, On positive invariance and output feedback stabilization of input constrained linear systems, in *American Control Conference*, vol. 3 (1994), pp. 2740–2744
9. F. Mesquine, A. Benzaouia, Existence of output feedback for the regulator problem of a class of systems with constrained control, in *1th International Conference on Electronics and Automatic Control*, Vol. 4 (Tizi-Ouzzou, Algeria, 1992)
10. J. Daafouz, P. Riedinger, C. Iung, Stability analysis and control synthesis for switched systems: a switched Lyapunov function approach. IEEE Trans. Aut. Control **47**(11), 1883–1887 (2002)
11. D. Dai, T. Hu, A.R. Teel, L. Zaccarian, Piecewise-quadratic Lyapunov functions for systems with deadzones or saturation. Syst. Control Lett. **58**(5), 365–371 (2009)
12. D. Dai, T. Hu, A.R. Teel, L. Zaccarian, Output feedback synthesis for sampled-data system with input saturation, in *American Control Conference, Baltimore, MD, USA* (2010)

13. H. Hindi, S. Boyd, Analysis of linear systems with saturating using convex optimization, in *Proceeding of the 37th IEEE CDC, Florida* (1998), pp. 903–908
14. T. Hu, Z. Lin, B.M. Chen, Analysis and design for discrete-time linear systems subject to actuator saturation. Syst. Control Lett. **45**, 97–112 (2002)
15. Y.Y. Cao, Z. Lin, Stability analysis of discrete-time systems with actuator saturation by saturation dependent Lyapunov function, in *Proceedings of the 41th IEEE CDC, Las Vegas, USA* (2002)
16. A. Benzaouia, F. Mesquine, A. Hmamed, H. Aoufoussi, Stability and Control Synthesis for Discrete-time Linear Systems Subject to Actuator Saturation by output feedback. Math. Prob. Eng. J. (2006)

Chapter 8
Stabilization of Unsymmetrical Saturated Control Systems

8.1 Introduction

Stabilization problem for linear systems with asymmetric constraints on the control is considered in this chapter. The objective here is an attempt to solve the stabilization problem for asymmetrically constrained systems but in the framework of LMIs. It is worth recalling that the positive invariance approach may handle nonsymmetrical constraints but conditions cannot be written easily under LMIs. In fact, results obtained in [6] and presented in chapter IV are expressed as LMIs but involve an equality LMI, for the pole assignment algebraic equation, that augment the difficulty to obtain feasible solutions. On the other side, results of the approach based on the convex writing of the constrained system allowing saturation to take effect developed by [10, 12, 13] are expressed under LMIs formalism but cannot handle unsymmetrical constraints.

It is well known that some works using constraints of symmetric nature as in [4, 9, 11–13] are expressed under LMI form. The characterization of invariance with symmetric convex sets for saturated systems is also used in the literature [8]. However, the asymmetric character of the actuator constraints is very important in practical situations since these constraints are inherently asymmetric. Many attempts were developed to emphasis LMIs and problems with asymmetric saturation but lead to conservative results as in [2, 3]. In this chapter, we address the regulator problem for discrete-time and continuous-time linear system with asymmetric saturation on the control in terms of an LMI problem. The main contribution of this work is to extend the results of [9], easily written under LMIs but restricted to symmetric constraints, to systems with asymmetric constraints and to be formulated under LMIs form obtained for the first time in [5].

A. Benzaouia et al., *Saturated Control of Linear Systems*, Studies in Systems, Decision and Control 124, https://doi.org/10.1007/978-3-319-65990-9_8

8.2 Stabilization by Unsymmetrical Constrained State Feedback Control: Discrete-Time Case

8.2.1 Problem Formulation

Considering a discrete-time system with a control that has an unsymmetrical saturation, the state equation is written as follows:

$$\begin{aligned} x(k+1) &= Ax(k) + BSat(u(k)), \\ x(0) &= x_o \end{aligned} \tag{8.1}$$

where $x(k) \in \mathbb{R}^n$ is the state vector and $u(k) \in \mathbb{R}^m$ is the control. In this section, the following assumption is necessary:

$$\text{A} \quad \text{is non singular} \tag{8.2}$$

The saturation of the control is nonsymmetrical, and the expression of each component of the vector $Sat(u)$ can be described by:

$$(Sat(u))_i = Sat(u_i) = \begin{cases} \alpha_i & \text{if} \quad u_i \geq \alpha_i \\ u_i & \text{if} \; -\beta_i \leq u_i \leq \alpha_i \\ -\beta_i & \text{if} \quad u_i \leq -\beta_i \end{cases} \tag{8.3}$$
$$\text{for} \;\; i = 1, \ldots, m \;\; \text{with} \;\; \alpha_i \neq \beta_i, \;\alpha_i > 0, \beta_i > 0$$

We use a state feedback control of type:

$$u(k) = LFx(k) + F_o \tag{8.4}$$

The nonsymmetrical saturation (8.3) together with the state feedback (8.4) induces the following set in the state space:

$$\mathcal{L}_{ns}(F) = \left\{x \in \mathbb{R}^n | - \Gamma e \leq LFx + F_o \leq \Lambda e\right\} \tag{8.5}$$

where matrices Λ, Γ, and the vector $e \in \mathbb{R}^m$ are given by:

$$\Lambda = \begin{bmatrix} \alpha_1 & 0 & \ldots & 0 \\ * & \alpha_2 & \ldots & 0 \\ * & * & \ldots & 0 \\ * & * & * & \alpha_m \end{bmatrix},$$
$$\Gamma = \begin{bmatrix} \beta_1 & 0 & \ldots & 0 \\ * & \beta_2 & \ldots & 0 \\ * & * & \ldots & 0 \\ * & * & * & \beta_m \end{bmatrix}. \tag{8.6}$$

$$e = \begin{bmatrix} 1 \, 1 \ldots 1 \end{bmatrix}^T . \tag{8.7}$$

The gain F has to stabilize the system, while the gains F_o, L have to symmetrize the asymmetrical set of induced states $\mathscr{L}_{ns}(F)$ as it will be shown in the sequel. The problem studied thereafter is to stabilize by state feedback control the saturated system (8.1)–(8.3). It is a classical problem where the novelty is to handle unsymmetrical saturation on the control in the frame work of LMIs.

8.2.2 Symmetrization Technique

In this section, the cornerstone of developments allowing to transform the asymmetrical problem to a symmetrical one is presented. Further, the main lemma of the works [10, 11, 13] and recalled in Chap. 1 will be used to write the saturating system as a convex combination of $\eta = 2^m$ linear systems.

For each component of the control u_i, one can make the following change of variables:

$$w_i = u_i - \frac{\alpha_i - \beta_i}{2}, \tag{8.8}$$

with this, one can then rewrite the saturation of the control as

$$Sat(u_i) = sat_s(w_i) + \frac{\alpha_i - \beta_i}{2}, \tag{8.9}$$

where $sat_s(w_i)$ defined by:

$$sat_s(w_i) = \begin{cases} \frac{\alpha_i+\beta_i}{2} & \text{if} \quad w_i \geq \frac{\alpha_i+\beta_i}{2} \\ w_i & \text{if} \; -\frac{\alpha_i+\beta_i}{2} \leq w_i \leq \frac{\alpha_i+\beta_i}{2} \\ -\frac{\alpha_i+\beta_i}{2} & \text{if} \quad w_i \leq -\frac{\alpha_i+\beta_i}{2} \end{cases} \quad \text{for } i = 1, \ldots, m. \tag{8.10}$$

A second change of variable is used:

$$z_i = w_i \frac{2}{\alpha_i + \beta_i}, \tag{8.11}$$

with $sat(z_i)$ standing for the normalized symmetric saturation given by:

$$sat(z_i) = \begin{cases} 1 & \text{if} \;\; z_i \geq 1 \\ z_i & \text{if} \; -1 \leq z_i \leq 1 \\ -1 & \text{if} \;\; z_i \leq -1 \end{cases} \quad \text{for } i = 1, \ldots, m \tag{8.12}$$

Using the change of variables (8.8) and (8.11), one can rewrite u_i as follows:

$$u_i = \frac{\alpha_i + \beta_i}{2} z_i + \frac{\alpha_i - \beta_i}{2} \tag{8.13}$$

In matrix notation, the expression (8.13) can be written as:

$$u = \frac{\Lambda + \Gamma}{2} z + \frac{\Lambda - \Gamma}{2} e, \tag{8.14}$$

With relation (8.13), the expression of $Sat(u_i)$ given by (8.3) is related to $sat(z_i)$ given by (8.12) as proven in Lemma below:

Lemma 8.1 *The nonsymmetrical saturation $Sat(u)$ is linked to the normalized symmetric saturation by the following relation:*

$$Sat(u) = (\frac{\Lambda + \Gamma}{2}) sat(z) + (\frac{\Lambda - \Gamma}{2}) e \tag{8.15}$$

Proof Three cases can be examined for $i = 1, \ldots, m$:

- $z_i \geq 1$, according to (8.13) and bearing in mind that $\Lambda + \Gamma > 0$, one obtains $u_i \geq \alpha_i$. Hence, $Sat(u_i) = \alpha_i$ while $sat(z_i) = 1$, which satisfies (8.15).
- $z_i \leq -1$, following the same lines, one can conclude that $u_i \leq -\beta_i$. Hence, $Sat(u_i) = -\beta_i$ while $sat(z_i) = -1$, which confirms (8.15).
- $-1 \leq z_i \leq 1$: obtained obviously by similar reasoning. □

By introducing these notations in the state Eq. (8.1), the term $BSat(u)$ can be developed as follows:

$$\begin{aligned} BSat(u) &= B(\frac{\Lambda + \Gamma}{2}) sat(z) + B(\frac{\Lambda - \Gamma}{2}) e \\ &= \tilde{B} sat(z) + \tilde{\xi}, \end{aligned} \tag{8.16}$$

where matrix $\tilde{B}$ and vector $\tilde{\xi}$ are given by:

$$\tilde{B} = B(\frac{\Lambda + \Gamma}{2}) \text{ and } \tilde{\xi} = B(\frac{\Lambda - \Gamma}{2}) e. \tag{8.17}$$

By introducing (8.15) in the state Eq. (8.1), the term $BSat(u)$ can be developed as (8.16) and (8.17). With these notations, we can rewrite the state equation of the nonsymmetrical system as follows:

$$x(k+1) = Ax(k) + \tilde{B} sat(z(k)) + \tilde{\xi}. \tag{8.18}$$

This system can be seen as a symmetrical saturated system with a bounded disturbance. The link between (8.4) and (8.14) is given by the following lemma.

Lemma 8.2 *The control (8.4) with* $L = \frac{\Lambda+\Gamma}{2}$ *and* $F_o = \frac{\Lambda-\Gamma}{2}e$ *symmetrizes the asymmetrical set* $\mathscr{L}_{ns}(F)$ *given by (8.5).*

Proof Let $x \in \mathscr{L}_{ns}(F)$, that is, $-\Gamma e \leq LFx + F_o \leq \Lambda e$. Substituting matrices L, F_o, one has $-\Gamma e - (\frac{\Lambda-\Gamma}{2})e \leq (\frac{\Lambda+\Gamma}{2})Fx \leq \Lambda e - (\frac{\Lambda-\Gamma}{2})e$. This inequality is equivalent to: $-(\frac{\Lambda+\Gamma}{2})e \leq (\frac{\Lambda+\Gamma}{2})Fx \leq (\frac{\Lambda+\Gamma}{2})e$. Noting that matrix $\frac{\Lambda+\Gamma}{2}$ is positive diagonal, we finally obtain: $-1 \leq Fx \leq 1$. □

The objective of this work is to design the gain F for the unsymmetrical saturated controller. Define the following sets:

$$\mathscr{L}(F) = \{x \in \mathbb{R}^n | \; |Fx|_i \leq 1, i = 1, \ldots, m\} \tag{8.19}$$

$$\varepsilon(P, \rho) = \{x \in \mathbb{R}^n | \; x^T Px \leq 1\} \tag{8.20}$$

$$\varepsilon_{nc}(P, \rho) = \left\{x \in \mathbb{R}^n | \; (x + \tilde{\xi})^T P(x + \tilde{\xi}) \leq 1\right\} \tag{8.21}$$

Henceforth, the gain feedback we are looking for will be designed to stabilize the system (8.18). Note that stabilizing these systems (symmetrical ones), one has to design a control with: $sat(z_k) = \sum_{s=1}^{\eta} \gamma_s(k)(D_s z_k + D_s^- v_k)$, with $z = Fx$ and $v = Hx$ the auxiliary control where matrices F and H are to be computed. Thus, for the stabilization problem by state feedback control, we have to determine two gain matrices F and H.

The system equation with saturation in closed loop, using Lemma 1.3, is then written as follows:

$$x(k+1) = Ax(k) + \tilde{B}\sum_{s=1}^{\eta} \gamma_s(k)(D_s F + D_s^- H)x(k) + \tilde{\xi},$$

or in the equivalent form:

$$x(k+1) = \sum_{s=1}^{\eta} \gamma_s(k) A_s^{cl} x(k) + \tilde{\xi} = A_{cl}x(k) + \tilde{\xi}, \tag{8.22}$$

where the matrix in closed loop A_{cl} is given by:

$$A_{cl} = \sum_{s=1}^{\eta} \gamma_s(k) A_s^{cl}$$

$$A_s^{cl} = A + \tilde{B}(D_s F + D_s^- H). \tag{8.23}$$

Notice that the set $\mathscr{L}(H)$ is defined by the same expression (8.19) of $\mathscr{L}(F)$ as well as $\mathscr{L}_{ns}(H)$ is defined by the same expression (8.5) of $\mathscr{L}_{ns}(F)$.

8.2.3 Constrained Control for Discrete-Time Systems

With the equivalent writing of the unsymmetrically saturated system in closed loop under symmetrical form developed above, we are able to derive sufficient conditions of stabilizability by using LMIs. Let us introduce the following notation:

$$\tilde{x}(k) = x(k) + \xi. \tag{8.24}$$

where $\xi = A^{-1}\tilde{\xi}$, bearing in mind that the system verifies the assumption (8.2). Using the changes of variable presented above, the state equation can be rewritten as follows:

$$\begin{aligned} \tilde{x}(k+1) &= A\tilde{x}(k) + \tilde{B}sat(z(k)) \\ \tilde{x}_0 &= x_o + \xi, \end{aligned} \tag{8.25}$$

if one uses a state feedback control:

$$z(k) = F\tilde{x}(k), \tag{8.26}$$

the system equation with saturation in closed loop, using Lemma 1.3, is then written as:

$$\tilde{x}(k+1) = A\tilde{x}(k) + \tilde{B}\sum_{s=1}^{\eta} \gamma_s(k)(D_s F + D_s^- H)\tilde{x}(k),$$

or in the equivalent form:

$$\tilde{x}(k+1) = \sum_{s=1}^{\eta} \gamma_s(k) A_s^{cl}\tilde{x}(k) = A_{cl}\tilde{x}(k), \tag{8.27}$$

where the matrix in closed-loop A_{cl} is given by (8.23).

The following theorem gives sufficient conditions for the system (8.1) with state feedback to be asymptotically stable.

Theorem 8.1 *If there exist matrices $H \in \mathbb{R}^{m\times n}$, $F \in \mathbb{R}^{m\times n}$ and a symmetric positive definite matrix $P \in \mathbb{R}^{n\times n}$, such that:*

$$A_{cl}^T P A_{cl} - P \prec 0, \tag{8.28}$$

$$\varepsilon(P,\rho) \subset \mathscr{L}(H), \tag{8.29}$$

then the system defined by (8.27) with $z = F\tilde{x}$ is asymptotically stable $\forall \tilde{x}_0 \in \varepsilon(P,\rho)$.

Proof Assuming that condition (8.29) holds true, the saturated system (8.1) can be written as given by (8.27). Consider the Lyapunov function candidate $V(\tilde{x}(k))$ which is defined by:

$$V(\tilde{x}(k)) = \tilde{x}^T(k)P\tilde{x}(k) \tag{8.30}$$

Its rate of increase $\triangle V(\tilde{x}(k))$, in closed loop, is given by:

$$\begin{aligned} \triangle V(\tilde{x}(k)) &= \tilde{x}^T(k+1)P\tilde{x}(k+1) - \tilde{x}^T(k)P\tilde{x}(k) \\ &= \tilde{x}^T(k)\left[A_{cl}^T\, P\, A_{cl} \;-\; P\right]\tilde{x}(k) \end{aligned} \tag{8.31}$$

It is clear that if condition (8.28) is satisfied, then the rate of increase (8.31) is negative. Therefore, the symmetrically saturated closed-loop system is asymptotically stable. Hence, the unsymmetrically saturated closed-loop system converges toward $-\xi$. □

The previous result gives sufficient conditions of stabilizability for the closed-loop system. Below, we reformulate these conditions under LMIs allowing to deduce easily the controller gains.

Corollary 8.1 *If there exist matrices Y, Z, and $X = X^T \succ 0$ such that the following LMIs are satisfied:*

$$\begin{bmatrix} X & (AX + \tilde{B}D_sY + \tilde{B}D_s^-Z)^T \\ * & X \end{bmatrix} \succ 0, \;\; s = 1, \ldots, \eta, \tag{8.32}$$

$$\begin{bmatrix} \mu & Z_i \\ * & X \end{bmatrix} \succ 0, \;\; i = 1, \ldots, m,$$

then the system (8.27) is asymptotically stable $\forall \tilde{x}_o \in \varepsilon(P, \rho)$, with $\mu = 1/\rho$, Z_i is the ith row of matrix Z. The controller gains that stabilize the system are as follows:

$$F = YX^{-1}, H = ZX^{-1}, \;\; P = X^{-1}, \tag{8.33}$$

Proof The sufficient condition of asymptotic stability of the saturated system is given by (8.28). Multiplying the left and right by $P^{-1} = X$ leads to: $X - (XA_{cl}^T)X^{-1}(A_{cl}X) \succ 0$. Using the Schur complement, we obtain:

$$\begin{bmatrix} X & (A_{cl}X)^T \\ * & X \end{bmatrix} \succ 0$$

By replacing the matrix of the closed-loop system A_{cl} by its expression (8.23) and using the change of variables $Y = FX^{-1}$ and $Z = HX^{-1}$, it is easy to obtain the LMIs (8.32). Consequently, the sufficient conditions of asymptotic stability (8.28), for the closed-loop system and for any initial state within the set $\varepsilon(P, \rho)$, are obtained.

Furthermore, the inclusion (8.29) is equivalent to $\rho H_i P^{-1} H_i^T \leq 1, i = 1, \ldots, m$. Develop equivalently as follows: $\rho(HX)_i X^{-1}(HX)_i^T \leq 1,\; i = 1, \ldots, m$, which is equivalent to $\rho Z_i X^{-1} Z_i^T \leq 1, i = 1, \ldots, m$. Using the Schur complement, we obtain the LMIs (8.33). Finally, using the change of variables (8.33), the LMIs (8.32)–(8.33) are deduced. To complete the proof, note that as the system (8.27) is asymptotically stable, the system (8.1) converges toward the point $-\xi$. □

Remark 8.1 The change of variables that symmetrizes the nonsymmetrical saturation introduces a closed-loop system that is affine. That is why the nonsymmetrical closed-loop system converges toward a value, namely $-\xi$ which is, in general, different of 0.

In the following example, we illustrate the obtained results.

Example 8.1 For this example, we have $n = 2$ and $m = 1$ and the bounds of the control are $\alpha = 5$ and $\beta = 10$. Consider the system governed by (8.25) with the following matrices:

$$A = \begin{bmatrix} 1 & 2 \\ 3 & 4 \end{bmatrix}; B = \begin{bmatrix} 2 \\ 1 \end{bmatrix}.$$

The LMIs (8.32) and (8.33) are feasible. The solution obtained for $\rho = 1$ is:

$$X = \begin{bmatrix} 115.6386 & -91.5921 \\ -91.5921 & 89.4067 \end{bmatrix};$$
$$Y = \begin{bmatrix} 3.8150 & -7.6789 \end{bmatrix};$$
$$Z = \begin{bmatrix} 3.8121 & -7.6756 \end{bmatrix};$$

and thus gains F and H for the closed-loop system, with a nonsymmetrical saturated control, are: $F = [-0.1858 \quad -0.2762]$; $H = [0.4644 \quad 0.6904]$.

Matrices L and F_o which are involved in the expression of control, given by Eq. (8.4), are as follows:

$$L = ((\Lambda + \Gamma)/2)e = 7.5,$$
$$F_o = ((\Lambda - \Gamma)/2)e = -2.5.$$

In Fig. 8.1, we present the stability ellipsoid, the set $\mathscr{L}_{ns}(H)$, and some trajectories of the state vector x. Figure 8.2 represents the control evolution with respect to time for one of the simulated trajectories.

8.3 Stabilization by Unsymmetrical Constrained State Feedback Control: Continuous-Time Case

8.3.1 Problem Formulation

The constrained studied system is given by:

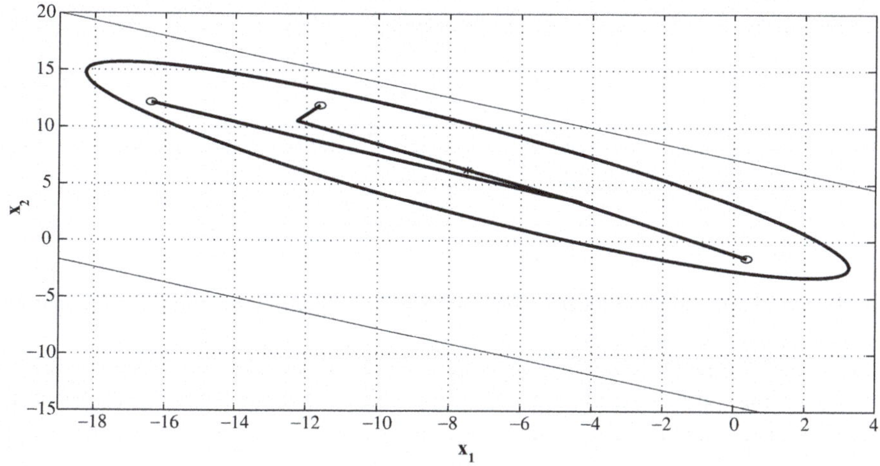

Fig. 8.1 Ellipse area of stability with $\mathscr{L}_{ns}(H)$ and some trajectories of the state vector x (o indicates the initial state)

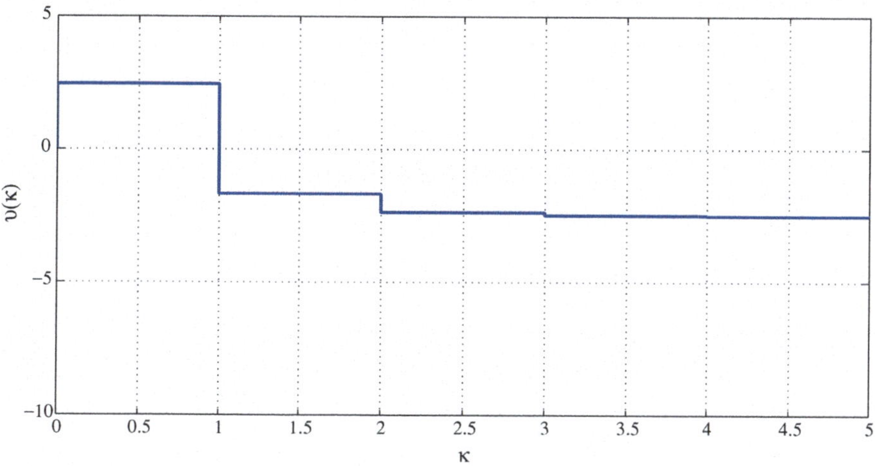

Fig. 8.2 Control evolution for one of the trajectories above

$$
\begin{aligned}
\dot{x}(t) &= Ax(t) + BSat(u(t)), \\
x_o &= x(0),
\end{aligned}
\tag{8.34}
$$

where the control u is nonsymmetrically saturated. In this section, the assumption (8.2) is necessary. The expression of each component of the vector $Sat(u)$ can be described by (8.3).

The control considered is the state feedback given by:

$$u(t) = Kx(t) + \omega \tag{8.35}$$

and defining as above diagonal matrices Λ and Γ. The set of induced constraints on the state space is given by:

$$\mathscr{L}_{ns}(K) = \left\{x \in \mathbb{R}^n | -\Gamma e \leq Kx + \omega \leq \Lambda e\right\} \tag{8.36}$$

Matrix K and vector ω will be designed such that the unsymmetrical saturation problem is transformed to a symmetrical one.

Hence, we are addressing the problem of giving under LMI formulation a design procedure enabling to deduce the state feedback K that stabilizes the closed-loop system with asymmetric constraints.

8.3.2 LMI Constrained Control

Similar change of variables used in the discrete-time case leads to the following development. With these notations, we can rewrite the state equation of the system as follows:

$$\dot{x}(t) = Ax(t) + \tilde{B}sat(z(t)) + \tilde{\xi} \tag{8.37}$$

To eliminate the constant vector $\tilde{\xi}$, a new state vector given by (8.24) is introduced, provided assumption, (8.2). The state equation becomes:

$$\dot{\tilde{x}}(t) = A\tilde{x}(t) + \tilde{B}sat(z(t)) \tag{8.38}$$

$$\tilde{x}_0 = x_o + \tilde{\xi}. \tag{8.39}$$

By using a state feedback control as:

$$z(t) = \tilde{K}\tilde{x}(t), \tag{8.40}$$

the system in closed loop, while the saturation does not occur, is given by:

$$\begin{aligned} \dot{\tilde{x}}(t) &= A_{cl}\tilde{x}(t), \\ A_{cl} &= (A + \tilde{B}\tilde{K}) \end{aligned} \tag{8.41}$$

The link between the effective gain control K and the gain controller of the transformed system (8.38) $\tilde{K}$ is given by:

$$K = \frac{\Lambda + \Gamma}{2}\tilde{K} \tag{8.42}$$

$$\omega = \frac{\Lambda + \Gamma}{2}\tilde{K}\tilde{\xi} + \frac{\Lambda - \Gamma}{2}e \tag{8.43}$$

It is worth noting that matrix A_{cl} can also be expressed as $A_{cl} = A + BK$.
Let us define the ellipsoid in the state space for the original system (8.34) as follows:

$$\varepsilon_{nc}(P, \rho) = \left\{x \in \mathbb{R}^n |\ (x + \tilde{\xi})^T P (x + \tilde{\xi}) \leq \rho\right\} \tag{8.44}$$

and for the system (8.41) as:

$$\varepsilon(P, \rho) = \left\{\tilde{x} \in \mathbb{R}^n |\ \tilde{x}^T P \tilde{x} \leq \rho\right\} \tag{8.45}$$

Further, the induced set, by state feedback, in the state space for system (8.41) is given by:

$$\mathcal{L}(\tilde{K}) = \left\{\tilde{x} \in \mathbb{R}^n |\ |\tilde{K}\tilde{x}|_i \leq 1, i = 1, \ldots, m\right\} \tag{8.46}$$

Sufficient conditions of stabilizability for systems with unsymmetrical constrained control are presented in this section. The control taken here is a state feedback given by (8.35).

The link between the symmetrical set $\mathcal{L}(\tilde{K})$ and the unsymmetrical set $\mathcal{L}_{ns}(K)$ defined by (8.36) is shown by the following Lemma:

Lemma 8.3 *For all $x \in \mathbb{R}^n$, the two following statements are equivalent:*

$$(i) \quad x \in \mathcal{L}_{ns}(K),$$
$$(ii) \quad \tilde{x} \in \mathcal{L}(\tilde{K}).$$

Proof ($(i) \Rightarrow (ii)$): Assume that $x \in \mathcal{L}_{ns}(K)$. The unsymmetrical set $\mathcal{L}_{ns}(K)$ defined by (8.36) is shown by the following relation:

$$-\Gamma e \leq Kx + \omega \leq \Lambda e \tag{8.47}$$

Using relations (8.42)–(8.43) and the fact that matrix $\frac{\Lambda+\Gamma}{2}$ is diagonal positive, one can write:

$$\begin{aligned} &\left[\frac{\Lambda + \Gamma}{2}\right]^{-1}\left[-\Gamma e - \omega + (\frac{\Lambda + \Gamma}{2})\tilde{K}\tilde{\xi}\right] \\ &\leq \tilde{K}\tilde{x} \leq \left[\frac{\Lambda + \Gamma}{2}\right]^{-1}\left[\Lambda e - \omega + \frac{\Lambda + \Gamma}{2}\tilde{K}\tilde{\xi}\right] \end{aligned} \tag{8.48}$$

Replacing ω by its expression given by (8.43), one obtains:

$$-e \le \tilde{K}\tilde{x} \le e, \tag{8.49}$$

which is exactly the set $\mathcal{L}(\tilde{K})$.

($(ii) \Rightarrow (i)$): Now assume that $\tilde{x} \in \mathcal{L}(\tilde{K})$. Consider the double inequality (8.49) and replace matrix $\tilde{K}$ and vector $\tilde{x}$ by their expressions given by (8.42) and (8.24). It follows:

$$-e \le \tilde{K}\tilde{x} \le e, \tag{8.50}$$

$$-\frac{\Lambda+\Gamma}{2}e - K\tilde{\xi} \le Kx \le \frac{\Lambda+\Gamma}{2}e - K\tilde{\xi} \tag{8.51}$$

By adding and subtracting $\frac{\Lambda+\Gamma}{2}e$ in both sides of inequality (8.50), and using the expression of ω given by (8.43), one obtains successively:

$$-\frac{\Lambda+\Gamma}{2}e + \frac{\Lambda-\Gamma}{2}e - \omega \le Kx \le \frac{\Lambda+\Gamma}{2}e + \frac{\Lambda-\Gamma}{2}e - \omega \tag{8.52}$$

$$-\Gamma e \le Kx + \omega \le \Lambda e, \tag{8.53}$$

which is the unsymmetrical set $\mathcal{L}_{ns}(K)$. □

Lemma 8.4 *For all $x \in \mathbb{R}^n$, the two following statements are equivalent:*

(i) $\varepsilon_{nc}(P, \rho) \subset \mathcal{L}_{ns}(K)$,
(ii) $\varepsilon(P, \rho) \subset \mathcal{L}(\tilde{K})$.

Proof $(i) \Rightarrow (ii)$: Let $\tilde{x} \in \varepsilon(P, \rho)$, that is $\tilde{x}^T P\tilde{x} \le \rho$. By using (8.24), one obtains $(x+\tilde{\xi})^T P(x+\tilde{\xi}) \le \rho$, and this implies that $x \in \varepsilon_{nc}(P, \rho)$. According to (i), $x \in \mathcal{L}_{ns}(K)$. By virtue of Lemma 8.3, $\tilde{x} \in \mathcal{L}(\tilde{K})$.

The reciprocal is obvious. □

The following theorem gives sufficient conditions for the system (8.38) to be asymptotically stable and equivalently the constrained system (8.34) to be convergent toward a new point.

Theorem 8.2 *If there exist matrix $K \in \mathbb{R}^{m\times n}$, and a symmetric positive definite matrix $P \in \mathbb{R}^{n\times n}$, such that:*

$$P(A+BK) + (A+BK)^T P \prec 0, \tag{8.54}$$

and

$$\varepsilon_{nc}(P, \rho) \subset \mathcal{L}_{ns}(K), \tag{8.55}$$

then the system defined by (8.34) with $u = Kx + \omega$ converges toward the equilibrium point $-\tilde{\xi}$, $\forall x_o \in \varepsilon_{nc}(P, \rho)$.

Proof Assume that condition (8.55) holds true. For any initial condition $x(0) \in \varepsilon_{nc}(P, \rho)$, according to Lemma (8.4), the inclusion $\varepsilon(P, \rho) \subset \mathscr{L}(\tilde{K})$ holds true. Hence, the saturated system (8.38)–(8.40) can be written as given by (8.41).

Consider the following Lyapunov function candidate $V(\tilde{x}(t))$:

$$V(\tilde{x}(t)) = \tilde{x}^T(t)P\tilde{x}(t) \tag{8.56}$$

Compute its derivative:

$$\dot{V}(\tilde{x}(t)) = \dot{\tilde{x}}^T P\tilde{x} + \tilde{x}^T P\dot{\tilde{x}} = \tilde{x}^T(A_{cl}^T P + PA_{cl})\tilde{x} \tag{8.57}$$

Remember that $A_{cl} = A + BK = A + \tilde{B}\tilde{K}$. It is clear that if (8.54) is satisfied, then it is a sufficient condition for the derivative (8.57) to be negative definite. Therefore, the symmetrically constrained closed-loop system is asymptotically stable [9]. Consequently, the state of the original system (8.34) with the control law (8.35) converges toward $-\tilde{\xi}$. □

The previous result gives sufficient conditions for stabilizability for the closed-loop system. Below, we reformulate these conditions in the form of LMIs that allow to deduce the controller gain.

Corollary 8.2 *If there exist matrices Y and $X = X^T \succ 0$ such that the following LMIs are satisfied:*

$$[AX + BY]^{sym} \prec 0, \tag{8.58}$$

and

$$\begin{bmatrix} \mu_i & Y_i \\ & X \end{bmatrix} \succ 0 \ , i = 1, \ldots, m, \tag{8.59}$$

then the system (8.34) with the control law (8.35) converges toward $-\tilde{\xi}$, $\forall x_o \in \varepsilon_{nc}(P, \rho)$, with $\mu_i = \frac{(\alpha_i+\beta_i)^2}{4\rho}$, Y_i is the ith row of matrix Y. Further, the controller gain that stabilizes the system is given by:

$$K = YX^{-1} \text{ with } P = X^{-1}, \tag{8.60}$$

Proof The sufficient condition of asymptotic stability of the symmetrical constrained system is given by (8.54). Multiplying the left and right sides by $P^{-1} = X$ leads to:

$$X(A + BK)^T + (A + BK)X \prec 0.$$

Using the change of variables $Y = KX$, it is easy to obtain the LMI (8.58) which is equivalent to the sufficient condition of the asymptotic stability (8.54), for the closed-loop system (8.38), for any initial state within the set $\varepsilon(P, \rho)$. Consequently, the state of the original system (8.34) with the control law (8.35) converges toward $-\tilde{\xi}$.

Furthermore, by virtue of Lemma 8.4, the inclusion (8.55) is equivalent to the inclusion for symmetrical sets $\varepsilon(P, \rho) \subset \mathscr{L}(\tilde{K})$. Hence, using [7] one can write $\rho \tilde{K}_i P^{-1} \tilde{K}_i^T \leq 1,\ i = 1, \ldots, m$. According to (8.42), one has $K = \frac{\Lambda+\Gamma}{2}\tilde{K}$. That is, $\tilde{K} = (\frac{\Lambda+\Gamma}{2})^{-1} K$ and therefore, $\tilde{K}_i = (\frac{\alpha_i+\beta_i}{2})^{-1} K_i,\ i = 1, \ldots, m$, matrix $\frac{\Lambda+\Gamma}{2}$ being diagonal. This leads to $(K)_i X (K)_i^T \leq \mu_i,\ i = 1, \ldots, m$, with $(1/\mu_i) = \rho(\frac{\alpha_i+\beta_i}{2})^{-2}$. Develop now equivalently as follows:

$(KX)_i X^{-1} (KX)_i^T \leq \mu_i,\ i = 1, \ldots, m$, which is equivalent to $Y_i X^{-1} Y_i^T \leq \mu_i$, $i = 1, \ldots, m$. Using the Schur complement, we obtain the LMIs (8.59). □

Comments 8.1 *The derived LMIs (8.58)–(8.59) deal effectively with unsymmetrical saturation. This result is obtained for the first time in [5] reducing considerably the conservatism of the results of [9]. However, the advantage of this approach enabling one to use LMIs for unsymmetrical constraints introduces an equilibrium point different of the origin.*

In the following example, we illustrate the obtained results.

Example 8.2 Consider the system governed by (8.34) with the following matrices:

$$A = \begin{bmatrix} 1 & -0.5 \\ 1 & 3 \end{bmatrix};\ B = \begin{bmatrix} 1 \\ 2 \end{bmatrix};$$

we have $n = 2$, $m = 1$, and the constraints on the control are: $\alpha = 5$, $\beta = 10$. The change of variables given by (8.17) and the constant vector $\tilde{\xi}$ in (8.24) become:

$$\tilde{B} = \begin{bmatrix} 7.5 \\ 15 \end{bmatrix};\ \tilde{\xi} = \begin{bmatrix} -2.8571 \\ -0.7143 \end{bmatrix}$$

The LMIs (8.58) and (8.59) are feasible. The solution obtained for $\rho = 10$ is:

$$X = \begin{bmatrix} 0.3535 & 0.2025 \\ 0.2025 & 0.2315 \end{bmatrix};$$

$$Y = \begin{bmatrix} -0.3075 & -0.504 \end{bmatrix};$$

and thus

$$\tilde{K} = \begin{bmatrix} 0.1008 & -0.3785 \end{bmatrix},\ K = \begin{bmatrix} 0.7562 & -2.8386 \end{bmatrix}$$

In Fig. 8.3, we present the stability ellipsoid and some trajectories of the state vector x. One can notice that all the trajectories converge to the equilibrium point

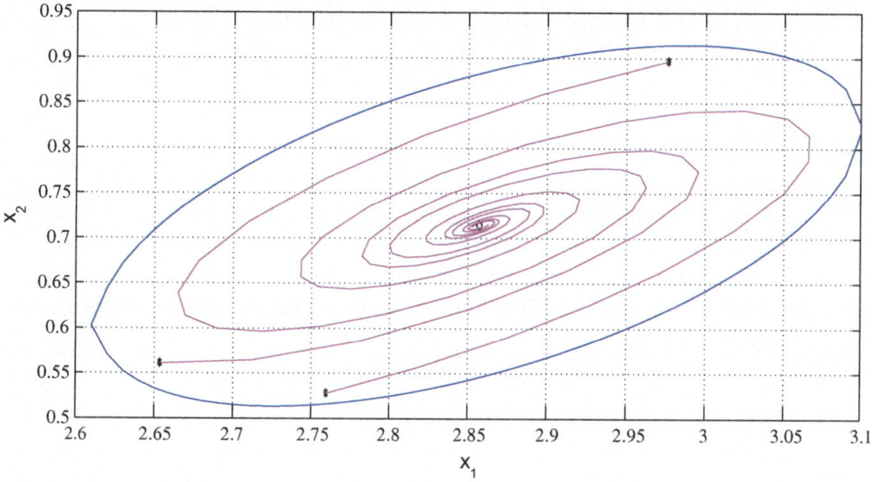

Fig. 8.3 Ellipsoid of stability and some trajectories of the state vector x

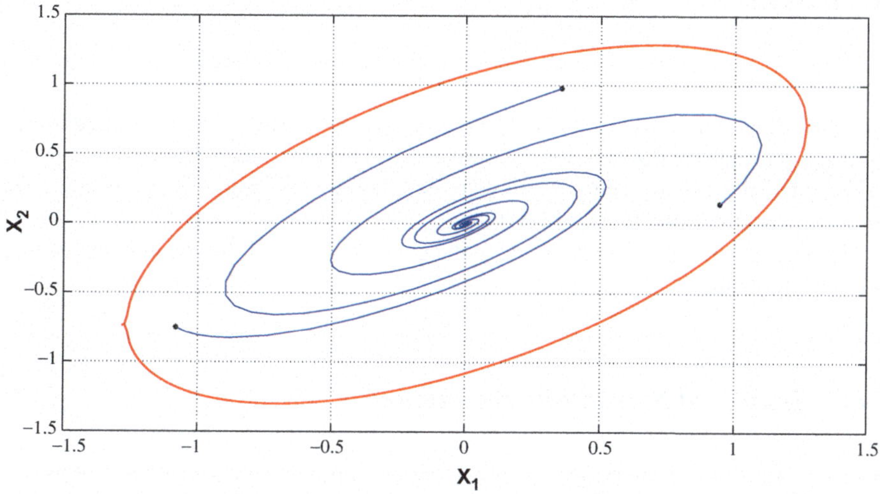

Fig. 8.4 Ellipsoid of stability and some trajectories of the state vector $\tilde{x}$

given by $-\tilde{\xi} = [2.8571 \;\; 0.7143]^T$. Further, it is worth noting that all the trajectories remain inside the ellipsoid $\varepsilon_{nc}(P, \rho)$. Figure 8.4 shows the asymptotic stability of the system (8.38) and some motions evolution inside the ellipsoid of stability $\varepsilon(P, \rho)$ from different initial conditions.

8.4 Constrained Control for Continuous-Time Case System: An Improved Technique

8.4.1 Problem Formulation

To overcome the assumption (8.2) and to use a new, less conservative, convex writing of the saturated system, the following approach may be of interest. The saturated studied system is given by (8.34). The expression of each component of the vector $Sat(u)$ can be described by (8.3). To stabilize the unsymmetrically saturated system, a state feedback control of type:

$$u(t) = LKx(t) + K_o, \tag{8.61}$$

is used.

The gain K has to stabilize the system, while the gains K_o, L play the role of symmetrizing the asymmetrical set $\mathscr{L}_{ns}(K)$ induced in the state space by the constraints and given as follows:

$$\mathscr{L}_{ns}(K) = \left\{x \in \mathbb{R}^n | -\Gamma e \leq LKx + K_o \leq \Lambda e\right\} \tag{8.62}$$

where the diagonal matrices Λ and Γ are given by (8.6). The problem studied thereafter is to stabilize by state feedback control (8.61) the saturated system (8.34), (8.3). It is a classical problem where the novelty is to handle unsymmetrical saturation on the control in the frame work of LMIs.

The objective of this work is to design the gains $K,\quad L,\quad K_o$ for the unsymmetrical saturated controller.

8.4.2 Improved Saturation Technique

In this section, the cornerstone of developments allowing to transform the asymmetrical problem to a symmetrical one is presented. Further, the main lemma of the work [17] is recalled. This last enables to write a saturated system in closed loop, as a convex combination of 2^m linear systems:

Lemma 8.5 ([17]) *For all $z \in \mathbb{R}^m$ and $v \in \mathbb{R}^{\tilde{m}}$, $\tilde{m} = m2^{m-1}$ such that $|v_i| \leq 1, i = 1, \ldots, \tilde{m}$.:*

$$sat(z) \in co\{D_s z + \hat{D}_s^- v\},\ s \in [1, \eta] \tag{8.63}$$

where D_s are diagonal matrices with each element of the diagonal either 1 *or* 0*, $D_s + D_s^- = \mathbb{I}_m$, $\eta = 2^m$, and $\hat{D}_s^- \in \mathbb{R}^{m\times\tilde{m}}$ is defined by:*

$$\hat{D}_s^- = e_{f_m(s)} \otimes D_s^-, s \in [1, \eta] \tag{8.64}$$

and $e_{f_m(s)} \in \mathbb{R}^{1\times 2^{m-1}}$ *is the row vector with zeros except* 1 *in the position* $f_m(s)$ *which is defined by:*

$$f_m(s) = \begin{cases} f_m(s-1)+1, & D_s + D_j \neq \mathbb{I}_m, \ \forall j \in [1, s] \\ f_m(j), & D_s + D_j = \mathbb{I}_m, \ \exists j \in [1, s] \end{cases} \tag{8.65}$$

The Lemma 8.5 allows to rewrite the saturated control using an auxiliary control ν which satisfies $|\nu_i| \leq 1$. Hence, there exist scalars $\gamma_s \geq 0$ $(s = 1, ..., \eta)$ with $\sum_{s=1}^{\eta} \gamma_s = 1$, such that:

$$sat(z(t)) = \sum_{s=1}^{\eta} \gamma_s(t)(D_s z(t) + \hat{D}_s^- \nu(t)) \tag{8.66}$$

The obtained closed-loop system becomes linear. On the other hand, for each component of the control u_i, one can make the change of variables as in (8.8), and with this change, one can then rewrite the saturation of the control as in (8.9), where *sats*(wi) is considered as the symmetrical non normalized saturation defined by (8.10).

A second change of variable is used as in (8.11), and let $sat(z_i)$ stands for the normalized symmetric saturation as in (8.12).

With the change of variables (8.8) and (8.11), one can rewrite u_i as (8.13). Or in matrix notation, the expression (8.13) can be written as:

$$u = \frac{\Lambda + \Gamma}{2} z + \frac{\Lambda - \Gamma}{2} e. \tag{8.67}$$

With relation (8.13), we prove in lemma below that the expression of $Sat(u_i)$ given by (8.3) is equivalent to $sat(z_i)$ given by (8.12) in expression (8.15).

By introducing (8.15) in the state Eq. (8.34), the term $BSat(u)$ can be developed as follows:

$$\begin{aligned} BSat(u) &= B(\frac{\Lambda + \Gamma}{2})sat(z) + B(\frac{\Lambda - \Gamma}{2})e \\ &= \tilde{B}sat(z) + Ew, \end{aligned} \tag{8.68}$$

where matrices E and $\tilde{B}$ are given by:

$$\tilde{B} = B(\frac{\Lambda + \Gamma}{2}), \quad E = \sqrt{n}B(\frac{\Lambda - \Gamma}{2}), w = \frac{e}{\sqrt{n}}. \tag{8.69}$$

With these notations, we can rewrite the state equation of the system as follows:

$$\dot{x}(t) = Ax(t) + \tilde{B}sat(z(t)) + Ew. \tag{8.70}$$

Note that $w^T w = 1$. In order to use available results on saturated systems, the obtained system (8.70), which is affine since w is known and constant, can be seen as a saturated one with a bounded disturbance.

Let us use a state feedback control of the form:

$$z(t) = Kx(t). \tag{8.71}$$

The link between control expression (8.4) and the one given by (8.67) is given by Lemma 8.2.

Define the following sets:

$$\mathscr{L}(K) = \{x \in \mathbb{R}^n |\ |Kx|_i \leq 1, i = 1, \ldots, m\} \tag{8.72}$$

$$\varepsilon(P, \rho) = \{x \in \mathbb{R}^n |\ x^T P x \leq \rho\} \tag{8.73}$$

Henceforth, for the stabilization problem, the system (8.70) is considered. Further, the gain feedback we are looking for will be designed to stabilize this system.

Note that stabilizing this system (symmetrical saturated system), one has to design a control using (8.66) with $z = Kx$ and $v = Hx$, $H \in \mathbb{R}^{\tilde{m} \times n}$ the auxiliary control with $|H_i x| \leq 1$, H_i the ith row of matrix H. The matrices K and H are to be designed.

The system equation with saturation in closed loop, using Lemma 8.5, is then written as follows:

$$\dot{x}(t) = Ax(t) + \tilde{B} \sum_{s=1}^{\eta} \gamma_s(t)(D_s K + \hat{D}_s^- H)x(t) + Ew, \tag{8.74}$$

or in the equivalent form:

$$\dot{x}(t) = \sum_{s=1}^{\eta} \gamma_s(t) A_s^{cl} x(t) + Ew = A_{cl} x(t) + Ew, \tag{8.75}$$

where the matrix in closed-loop A_c is given by:

$$\begin{aligned} A_{cl} &= \sum_{s=1}^{\eta} \gamma_s(t) A_s^{cl} \\ A_s^{cl} &= A + \tilde{B}(D_s K + \hat{D}_s^- H). \end{aligned} \tag{8.76}$$

Notice that the set $\mathscr{L}_{ns}(H)$ is defined by the same expression (8.36) of $\mathscr{L}_{ns}(K)$, while $\mathscr{L}(H)$ is defined by:

$$\mathscr{L}(H) = \{x \in \mathbb{R}^n |\ |Hx|_i \leq 1, i = 1, \ldots, \tilde{m}\} \tag{8.77}$$

8.4.3 LMI Constrained Control

The following theorem gives sufficient conditions for the system (8.74) to be strictly invariant in the sense of the following definition.

Definition 8.1 ([14]) A set in $\mathbb{R}^n$ is said to be invariant if all the trajectories starting from it will remain in it regardless of w.

An ellipsoid $\varepsilon(P, \rho)$ is said to be strictly invariant if $\dot{V} = 2x^T P(Bsat(Fx) + Ew) < 0$ for all w such that $w^T w \leq 1$ and all $x \in \partial\varepsilon(P, \rho)$, the boundary of $\varepsilon(P, \rho)$, where $V(x) = x^T Px$.

Theorem 8.3 *If there exist matrices $H \in \mathbb{R}^{\tilde{m}\times n}$, $K \in \mathbb{R}^{m\times n}$, a symmetric positive definite matrix $P \in \mathbb{R}^{n\times n}$, and positive scalars ρ, ς, such that:*

$$(A_s^{cl})^T P + PA_s^{cl} + \frac{1}{\varsigma}PEE^T P + \frac{\varsigma}{\rho}P \prec 0, \quad s = 1, \ldots, \eta, \tag{8.78}$$

and

$$\varepsilon(P, \rho) \subset \mathscr{L}(H), \tag{8.79}$$

where the matrix A_s^{cl} is given by (8.76), then the set $\varepsilon(P, \rho)$ is a strictly invariant set for system (8.75).

Proof The proof follows the same reasoning as the one of [14] where the classical convex writing of the saturation is replaced by the one given by Lemma 8.5. □

Similar result can be found in [16] where state constraints are also considered.

With the equivalent writing of the unsymmetrically saturated system in closed loop under symmetrical form developed above, we are able to derive sufficient conditions of stabilizability by using LMIs. The previous result gives sufficient conditions for stabilizability for the closed-loop system. Below, we reformulate these conditions in the form of LMIs that allow to deduce the controller gain.

Corollary 8.3 : *For positive scalars ρ, ς if there exist matrices $Z \in \mathbb{R}^{\tilde{m}\times n}$, $Y \in \mathbb{R}^{m\times n}$, and $X = X^T \in \mathbb{R}^{n\times n}$, $X \succ 0$ such that the following LMIs are satisfied:*

$$[AX + B(D_s Y + \hat{D}_s^- Z)]^{sym} + \frac{\varsigma}{\rho}X + \frac{1}{\varsigma}EE^T \prec 0, \; s = 1, \ldots, \eta, \tag{8.80}$$

$$\begin{bmatrix} \mu & Z_i \\ * & X \end{bmatrix} \succ 0, \; i = 1, \ldots, \tilde{m}, \tag{8.81}$$

with matrix D_s stands for a diagonal matrix with component either $\frac{\Lambda_s + \beta_s}{2}$ or 0, $D_s + D_s^- = \frac{\Lambda + \Gamma}{2}$ and $\hat{D}_s^-$ is defined by:

$$\hat{D}_s^- = e_{fm(s)} \otimes D_s^-, s \in [1, \eta], \tag{8.82}$$

then the set $\varepsilon(P, \rho)$ *is a strictly invariant set for system (8.75), with* $\mu = 1/\rho$, Z_i *is the* ith *line of matrix* Z. *The controller gains are given by:*

$$K = YX^{-1} \text{ and } H = ZX^{-1}, \text{ with } P = X^{-1}. \tag{8.83}$$

Proof The sufficient condition of invariance of the set $\varepsilon(P, \rho)$ with respect to the saturated system is given by (8.78). Using the Schur complement, one obtains

$$(A_s^{cl})^T P + PA_s^{cl} + \frac{\varsigma}{\rho}P + \frac{1}{\varsigma}PEE^T P \prec 0, \; s = 1, \ldots, \eta. \tag{8.84}$$

Multiplying the left and right sides of inequality (8.84) by $X = P^{-1}$ leads to LMIs (8.80) while replacing $\tilde{B}D$ by BD and using the change of variables $Y = KX^{-1}$, $Z = HX^{-1}$. These conditions are equivalent to the sufficient conditions of asymptotic stability (8.78), for the closed-loop system, for any initial state within the set $\varepsilon(P, \rho)$.

Furthermore, the inclusion (8.79) is equivalent to $\rho H_i P^{-1} H_i^T \leq 1, i = 1, \ldots, \tilde{m}$. Develop equivalently as follows:

$\rho(HX)_i X^{-1}(HX)_i^T \leq 1, i = 1, \ldots, \tilde{m}$, which is equivalent to $\rho Z_i X^{-1} Z_i^T \leq 1$, $i = 1, \ldots, \tilde{m}$. Using the Schur complement, we obtain the LMIs (8.81). □

Instead of using Lemma 8.5, one can use the convex writing of saturation given in [10, 11, 13],

$$sat(z) \in co\{D_s z + D_s^- v\}, \; s \in [1, \eta] \tag{8.85}$$

The closed-loop system becomes:

$$\dot{x}(t) = \sum_{s=1}^{\eta} \gamma_s(t) A_s^{cl} x(t) + Ew(t) = A_{cl}x(t) + Ew(t), \tag{8.86}$$

where the matrix in closed-loop A_s^{cl} is given by (8.76). In this case, Corollary 8.3 can be re-enunciated as:

Corollary 8.4 : *For positive scalars* ρ, ς *if there exist matrices* $Z \in \mathbb{R}^{m\times n}$, $Y \in \mathbb{R}^{m\times n}$, *and* $X = X^T \in \mathbb{R}^{n\times n}$, $X \succ 0$ *such that the following LMIs are satisfied:*

$$[AX + B(D_s Y + D_s^- Z)]^{sym} + \frac{\varsigma}{\rho}X + \frac{1}{\varsigma}EE^T \prec 0, \; s = 1, \ldots, \eta, \tag{8.87}$$

$$\begin{bmatrix} \mu & Z_i \\ * & X \end{bmatrix} \succ 0, \; i = 1, \ldots, m, \tag{8.88}$$

then the set $\varepsilon(P, \rho)$ *is a strictly invariant set for system (8.75), with* $\mu = 1/\rho$, Z_i *is the* ith *line of matrix* Z. *The controller gains that stabilize the system are as (8.83).*

Comments 8.2

- *It is worth noting that the convex expression (8.85) is more conservative than expression (8.63) for* $m > 1$, *according to [17]. In order to compare results obtained upon both expressions, Corollaries 8.3 and 8.4 are presented and tested in the example below.*
- *These LMIs are established by using the symmetric control* z. *However, by replacing matrix* $\tilde{B}$ *and* E *by their expressions with* Λ_i *and* Γ_i, *one take account of the asymmetry of the saturation on the control. Consequently, the derived LMIs (8.80)–(8.81) deal in reality with unsymmetrical saturation. This result is obtained for the first time in [5] reducing considerably the conservatism of the results of [17].*
- *The improved approach is presented for continuous-time systems only. The extension to the discrete-time systems can be easily obtained replacing the stability conditions by the discrete-time ones. The development concerning the saturation is being the same for both cases.*

In the following example, we illustrate the obtained results.

Example 8.3 : Consider the system governed by (8.34) with the following matrices:

$$A = \begin{bmatrix} -1 & 0.7 \\ 1 & 1 \end{bmatrix}; B = \begin{bmatrix} 1 & 0.2 \\ -0.3 & 0.5 \end{bmatrix} \tag{8.89}$$

For this example, we have $n = 2$, $m = 2$ and the control bounds are: $\alpha_1 = 5$, $\beta_1 = 10$, $\alpha_2 = 10$, and $\beta_2 = 5$.
It follows:

$$\tilde{B} = \begin{bmatrix} 7.5 & 1.5 \\ -2.25 & 3.75 \end{bmatrix},$$

$$E = \begin{bmatrix} -3.5355 & -0.7071 \\ -1.0607 & 1.7678 \end{bmatrix}$$

We solve LMIs (8.80) and (8.81). The obtained solutions in this case for $\rho = 1$ and $\varsigma = 1$ are:

$$X = \begin{bmatrix} 401.3375 & -161.6653 \\ -161.6653 & 67.4476 \end{bmatrix};$$

$$Y = \begin{bmatrix} 22.9276 & -20.2454 \\ -11.8235 & -35.2446 \end{bmatrix};$$

$$Z = \begin{bmatrix} 5.2535 & -3.0268 \\ -5.6885 & 0.9464 \\ 0.0496 & -0.0218 \\ -0.3650 & 0.1467 \end{bmatrix};$$

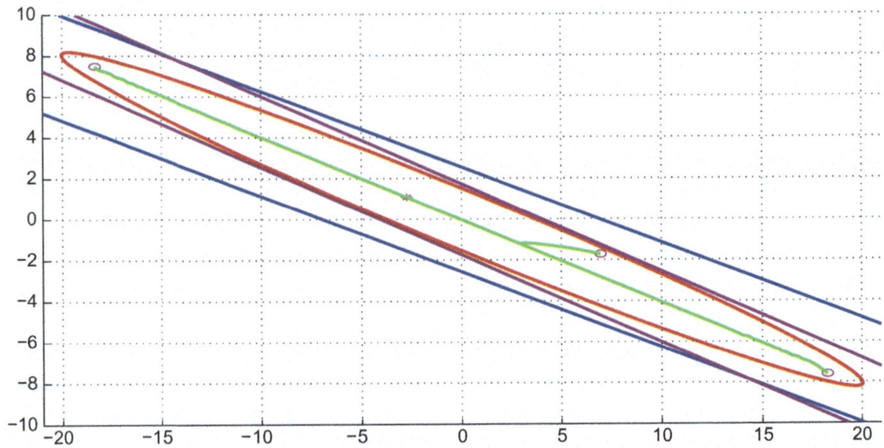

Fig. 8.5 Trajectories of the state vector x converging to the equilibrium point x_e

and thus gains K and H for the closed-loop system, with a nonsymmetrical saturated control, are:

$$K = \begin{bmatrix} -1.8494 & -4.7330 \\ -6.9574 & -17.1987 \end{bmatrix};$$

$$H = \begin{bmatrix} -0.1446 & -0.3915 \\ -0.2471 & -0.5782 \\ -0.0002 & -0.0008 \\ -0.0010 & -0.0001 \end{bmatrix}.$$

Figure 8.5 shows some trajectories of the state vector x with different initial states x_o. If $x_o \in \varepsilon_{nc}(P, \rho)$, then the trajectory converges surely to the equilibrium point given by $x_e = -(A + \tilde{B}K)^{-1}Ew$ which is closed to the origin due to the presence of the pseudo-permanent perturbation w. Figure 8.6 represents the inclusion of the ellipsoid set $\varepsilon(P, \rho)$ inside the polyhedral set of saturation $\mathscr{L}(H)$.

Example 8.4 In order to compare between Corollaries 8.3 and 8.4, system (8.89) is slightly modified as follows:

$$A = \begin{bmatrix} a & 0.7 \\ 1 & 1 \end{bmatrix}; B = \begin{bmatrix} b & 0.2 \\ -0.3 & 0.5 \end{bmatrix}$$

The feasibility of LMIs (8.87)–(8.88) and (8.80)–(8.81) is tested for a, b varying from -1 to 2 by a step of 0.1. The result of comparison is plotted in Fig. 8.7. Feasibility cases of (8.87)–(8.88) are indicated with $'+'$, while feasibility cases of (8.80)–(8.81) are indicated with $'\circ'$, showing the less conservatism of Corollary 8.3 based on the approach of [17].

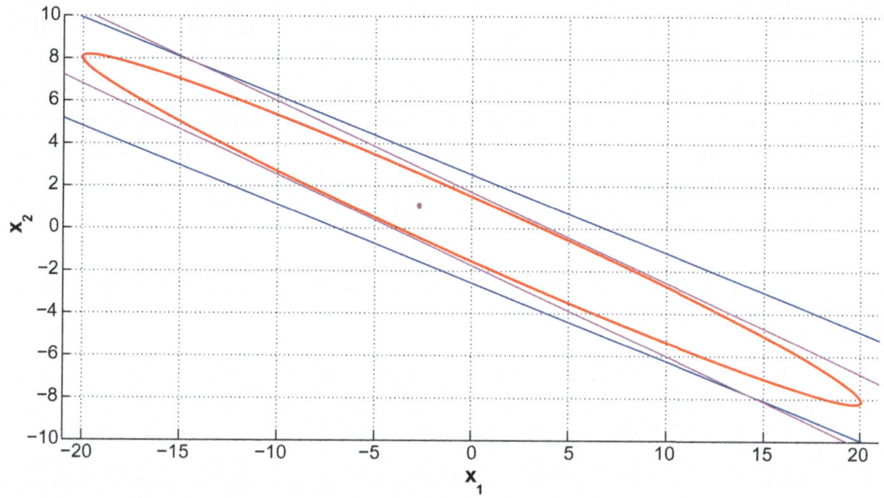

Fig. 8.6 Inclusion $\varepsilon(P, \rho) \subset \mathscr{L}(H)$ with the equilibrium point x_e

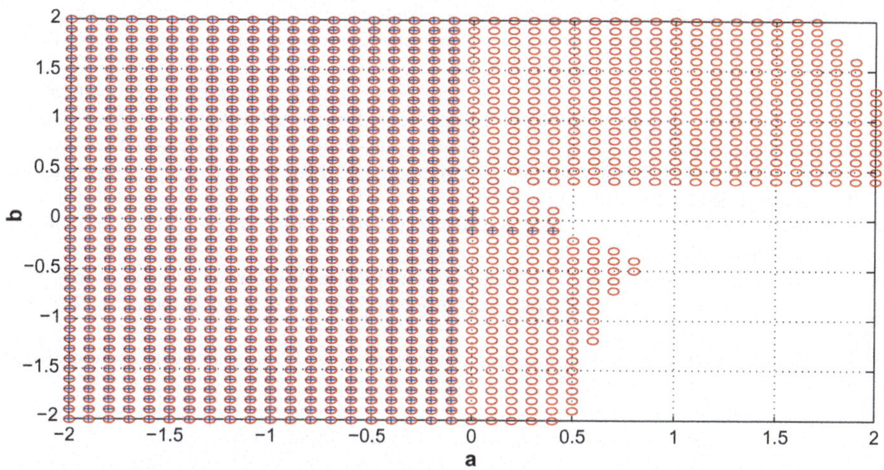

Fig. 8.7 Feasibility comparison

8.5 Conclusion

In this chapter, the regulator problem for linear (discrete-time or continuous-time) systems with asymmetric saturations on the control is developed in terms of an LMI problem. This work allows the results of [12], easily written under LMIs, to be extended to systems with asymmetric saturations formulated also under LMIs form. This was the main contribution of this chapter. The proposed approach first needs an assumption that the matrix system must be non-singular. This may be seen as a conservative assumption. However, a second enhanced approach is proposed to remove the restrictive assumption and leading to more general result for asymmetrical constraints in framework of LMIs.

Besides, these results extend those of the same authors developing unsaturating controllers working inside a region of linear behavior [1]. Numerical examples are studied to illustrate the proposed methodology and to show that the less conservative results are the ones based on the approach of [17].

References

1. M. Benhayoun, A. Benzaouia, F. Mesquine, A. El Hajjaji, System stabilization by unsymmetrical saturated state feedback control, in *Asian Conference of Control*, June 24–26, (Istanbul-Turkey, 2013)
2. A. Benzaouia, M. Ait Rami, S. El Faiz, Stabilization of linear systems with saturation: a Sylvester equation approach. J. Math. Control Inf. **21**(3), 247–259 (2004)
3. A. Benzaouia, S. El Faiz, The Regulator problem for linear systems with constrained control: an LMI approach. IMA J. Math. Control Inf. **23**(3), 335–345 (2006)
4. A. Benzaouia, F. Mesquine, A. Hmamed, H. Aoufoussi, Stability and control synthesis for discrete-time linear systems subject to actuator saturation by output feedback. Math. Probl. Eng. (2006)
5. A. Benzaouia, M. Benhayoun, F. Mesquine, Stabilization of systems with unsymmetrical saturated control: an LMI approach. Circuit Syst. Signal Process. **33**(10), 3263–3275 (2014)
6. A. Benzaouia, F. Tadeo, F. Mesquine, The regulator problem for linear systems with saturations on the control and its increments or rate: an LMI approach. IEEE Trans. Circuits Syst. I Fundam. Theory Appl. **53**(12) 2681–2691 (2006)
7. S. Boyd, L. EL Ghaoui, E. Feron, V. Balakrishnan, *Linear Matrix Inequalities in System and Control Theory* (SIAM, Philadelphia, PA, 1994)
8. M. Fiacchini, S. Tarbourich, C. Prieur, Invariance of symmetric convex sets for discrete-time saturated systems, in *50th IEEE CDC and European Control Conference* (Orlando, FL, USA, 2011)
9. P.O. Gutman, P. Hagander, A new design of constrained controllers for linear systems. IEEE Trans. Autom. Control **30**(1), 22–33 (1985)
10. T. Hu, Z. Lin, The equivalence of several set invariance conditions under saturation, in *Proceedings of the American Control Conference* (Chicago, Illinois, USA, 2000)
11. T. Hu, Z. Lin, The equivalence of several set invariance conditions under saturation, in *Proceedings of the 41th IEEE CDC* (Las Vegas, USA, 2002)
12. T. Hu, Z. Lin, B.M. Chen, Analysis and design for discrete-time linear systems subject to actuator saturation. Syst. Control Lett. **45**, 97–112 (2002)
13. T. Hu, Z. Lin, B.M. Chen, An analysis and design method for linear systems subject to actuator saturation and disturbance. Automatica **38**, 351–359 (2002)
14. T. Hu, Z. Lin, An analysis and design method for linear systems subject to actuator saturation and disturbance, in *Proceedings of the 41th IEEE CDC* (Las Vegas, USA, 2002)
15. F. Mesquine, F. Tadeo, A. Benzaouia, Regulator problem for linear systems with constraints on the control and its increments or rate. Automatica **40**(8), 1378–1395 (2004)
16. F. Mesquine, H. Ayad, M. Ait Rami, Disturbance attenuation for continuous time systems with state and control constraints, in *1st Conference on Systems and Control* (Marrakesh, 2007)
17. B. Zhou, Analysis and design of discrete-time linear systems with nested actuator saturation. Syst. Control Lett. **62**(10), 871–879 (2013)

Chapter 9
Delay Systems with Saturating Control

9.1 Introduction

It is clear that both input saturation and time delay can be encountered simultaneously in practical systems. In this case, the situation becomes very complicated. Many approaches were proposed to deal with this problem [3, 7, 14, 15, 18]. In the first part of this chapter, the same approach, of writing the saturated system [8, 10] as convex combination of linear delay systems, is applied. Partitioning of the delay interval is introduced in the Lyapunov–Krasovskii functional. The obtained LMIs are less conservative as the delay interval is partitioned onto r sub intervals [10] introducing more variables for the solution. As shown in [1], using the partition of the delay interval, some unfeasible problems become feasible. Further, a set of initial conditions that ensures the asymptotic stability of the closed-loop system is easily obtained from the solution to the proposed LMIs. Furthermore, the problem of guaranteed stability rate decay for the closed-loop system is also solved. Examples are presented to show the application and the less conservatism of the obtained conditions.

In the second part, the bound on the values that the delay can have is taken into account for stability and stabilizability conditions. The results obtained in the former part are delay independent and may be conservative when the bounds of the delay are known. Here, an improved delay dependent criteria [16] is extended to the delay saturating systems. Improved delay dependent criteria is first presented and then extended to the case of stabilizability condition to derive the controller [11]. Second, this condition is extended to saturating delay systems. Results are derived under LMI formalism to enable the synthesis of stabilizing memory-less state feedback. The obtained LMIs seem less conservative as the delay bound information is used. An illustrative example is presented and compared to previous results to show the effectiveness of the approach.

A. Benzaouia et al., *Saturated Control of Linear Systems*, Studies in Systems, Decision and Control 124, https://doi.org/10.1007/978-3-319-65990-9_9

9.2 Problem Statement

Let us consider the time-delay linear system with saturating control given by:

$$\begin{cases} \dot{x}(t) &= Ax(t) + A_\tau x(t-\tau) + Bsat(u(t)) \\ x(t) &= \Phi(t) \text{ for } t \in [-\tau, 0) \end{cases} \tag{9.1}$$

where $x(t)$ is the state of the system, $u(t)$ the control, τ the delay. The function $\Phi(t) \in \mathscr{C}_{n,\tau}$. The saturation is the standard non-linearity as:

$$sat(u_i(t)) = sign(u_i(t))\ min(1,\ |u_i(t)|) \tag{9.2}$$

Assume that one uses the memoryless state feedback given by:

$$u(t) = F\,x(t) \tag{9.3}$$

the closed-loop system becomes:

$$\begin{cases} \dot{x}(t) &= Ax(t) + A_\tau x(t-\tau) + Bsat(Fx(t)) \\ x(t) &= \Phi(t) \text{ for } t \in [-\tau, 0),\ \Phi(t) \in \mathscr{C}_{n,\tau} \end{cases} \tag{9.4}$$

For a given positive definite matrix P, we define the ellipsoid set:

$$\varepsilon(P,\ \rho) = \{x \in \mathbb{R}^n / x^T P x \leq \rho\} \tag{9.5}$$

The set of admissible controls where the linear behavior of the closed-loop system is guaranteed is given by $\mathscr{L}(F)$:

$$\mathscr{L}(F) = \{x \in \mathbb{R}^n / |F_i x| < 1\} \tag{9.6}$$

where F_i stands for the ith row of matrix F.

The domain of attraction to the origin in this case is given by all initial condition functions ψ such that:

$$\mathscr{S} = \{\psi\ \in \mathscr{C}_{n,\tau} /\ \text{the motion}\ \ x(t,\psi) \rightarrow 0 \text{ for } t \rightarrow \infty\} \tag{9.7}$$

Pb 1: For a given time-delay system (9.1), find memoryless state feedback F and a domain $\mathscr{D}$ of safe initial condition included in the domain of attraction of the system of the closed-loop system (9.4) such that asymptotic stability is guaranteed. Further, the problem of guaranteed rate decay α for the closed-loop system in the sense of definition given in (and recalled thereafter) [12] is also studied.

Pb 2: When the information about the value of the delay τ is available, the same problem is addressed with delay dependent conditions.

9.3 Partitioning for Stabilizability of Time-Delay Systems

For a vector $v \in \mathbb{R}^m$ such that $|v| < 1$, the saturation of vector $u \in \mathbb{R}^m$ can be written as a convex combination of two terms as given in Lemma 1.3 and hence, the closed-loop system can be re-written as ($\eta = 2^m$):

$$\begin{cases} \dot{x}(t) = Ax(t) + A_\tau x(t-\tau) + B \sum_{i=1}^{\eta} \gamma_i(t)(D_i F + D_i^- H)x(t) \\ x(t) = \Phi(t) \quad \text{for } t \in [-\tau, 0), \quad \Phi(t) \in \mathscr{C}_{n,\tau} \end{cases} \tag{9.8}$$

this enables one to recall the following sufficient conditions of stabilizability.

Theorem 9.1 ([2]) *Let the gain feedback F be given, for given $P \succ 0$, $Q \succ 0$, and $\rho > 0$ consider the set*

$$\mathscr{L}_v(\rho) = \{\phi \in \mathscr{C}_{n,\tau} / \phi(0)^T P \phi(0) + \int_{-\tau}^{0} \phi(\sigma)^T Q \phi(\sigma) d\sigma \leq \rho\} \tag{9.9}$$

if there exists a matrix $H \in \mathbb{R}^{m \times n}$ such that $\varepsilon(P, \rho) \subset \mathscr{L}(H)$ and:

$$\begin{bmatrix} (A_i^{cl})^T P + P(A_i^{cl}) + Q & PA_\tau \\ * & -Q \end{bmatrix} \prec 0, \tag{9.10}$$

where: $A_i^{cl} = A + B(D_i F + D_i^- H)$, for $i = 1, ..., \eta$, then the solution $x(t) \equiv 0$ is asymptotically stable and the set $\mathscr{L}_v(\rho)$ is an invariant set inside the domain of attraction.

It is worth to note here that the Lyapunov–Krasovskii approach is used and the proposed functional is given by:

$$V(x) = x^T(t)\ P\ x(t) + \int_{t-\tau}^{t} x(\sigma)^T\ Q\ x(\sigma)\ d\sigma \tag{9.11}$$

In what follows, less conservative delay independent stability condition based on partitioning the delay interval onto r equal sub intervals is proposed. In fact, for each interval a matrix Q_r can be found for the LMI (9.10) instead of using one constant matrix Q for all the delay interval which may be restrictive [1].

Proposition 9.1 *Let the gain feedback F be given, For given $P \succ 0$, $Q_k \succ 0$, $k = 1, \ldots r$ and $\rho > 0$, if there exists a matrix $H \in \mathbb{R}^{m \times n}$ such that:*

$$\begin{bmatrix} (A_i^{cl})^T P + P(A_i^{cl}) + Q_r & 0 & PA_\tau \\ 0 & \bar{Q} & 0 \\ * & 0 & -Q_1 \end{bmatrix} \prec 0 \tag{9.12}$$

$$\varepsilon(P,\rho) \subset \mathcal{L}(H) \tag{9.13}$$

where:

$$\bar{Q} = \begin{bmatrix} Q_1 - Q_2 & 0 & \dots & 0 \\ 0 & Q_2 - Q_3 & 0 \dots & 0 \\ & & \ddots & \\ 0 & \dots & 0 & Q_{r-1} - Q_r \end{bmatrix},$$

and r is the number of equal partitions of the interval $[t-\tau, t)$ then the solution $x(t) \equiv 0$ is asymptotically stable and the set:

$$\mathcal{L}_v^p(\rho) = \{\phi \in \mathcal{C}_{n,\tau} / \phi(0)^T P\phi(0) + \sum_{k=1}^{r} \int_{-\tau+(k-1)\frac{\tau}{r}}^{-\tau+k\frac{\tau}{r}} \phi(\sigma)^T Q_k \phi(\sigma) d\sigma \le \rho\} \tag{9.14}$$

is an invariant set inside the domain of attraction.

Proof For a state vector $x \in \varepsilon(P,\rho)$ assuming that $\varepsilon(P,\rho) \subset \mathcal{L}(H)$, it is possible to write

$$sat(Fx(t)) = \sum_{i=1}^{\eta} \gamma_i(t)(D_i F + D_i^- H)x(t). \tag{9.15}$$

Let

$$V(x) = x^T(t)\ P\ x(t) + \int_{t-\tau}^{t} x(\sigma)^T\ Q\ x(\sigma)\ d\sigma \tag{9.16}$$

be a candidate Lyapunov–Krasovskii function, using the partitioning of the interval $[t-\tau, t)$ such that matrix Q can be considered constant at each partition, $Q = Q_k, k = 1, ..., r$ for the interval k, noting $\tau_k = t - \tau + k\frac{\tau}{r}$, one can write:

$$V(x) = x^T(t)\ P\ x(t) + \sum_{k=1}^{r} \int_{\tau_{k-1}}^{\tau_k} x(\sigma)^T\ Q_k\ x(\sigma)\ d\sigma \tag{9.17}$$

First, we note that there exists non decreasing functions ε_1 and ε_2 such that

$$\varepsilon_1 \|x_t(0)\| \le V(x(t)) \le \varepsilon_2 \|x_t\| \tag{9.18}$$

where $\varepsilon_1 = \lambda_{min}(P)$ and $\varepsilon_2 = \lambda_{max}(P) + \tau \max_{k=1,\dots,r} \lambda_{max}(Q_k)$. On the other hand, the derivative of this functional along the trajectories of the system is as follows:

$$
\begin{aligned}
\dot{V}(x(t)) &= \dot{x}(t)^T Px(t) + x(t)^T P\dot{x}(t) + \sum_{k=1}^{r}[x(\tau_k)^T Q_k x(\tau_k) - x(\tau_{k-1})^T Q_k x(\tau_{k-1})] \\
&= x(t)^T(A^T + \left[B\sum_{i=1}^{\eta}\gamma_i(t)(D_i F + D_i^- H) \right]^T)Px(t) + x(t-\tau)^T A_\tau^T Px(t) \\
&+ x(t)^T P(A + \left[B\sum_{i=1}^{\eta}\gamma_i(t)(D_i F + D_i^- H) \right])x(t) + x(t)^T PA_\tau x(t-\tau) \\
&+ x(t)^T Q_r x(t) - x(t-\tau)^T Q_1 x(t-\tau) + x(\tau_1)^T Q_1 x(\tau_1) - x(\tau_{r-1})^T Q_r x(\tau_{r-1}) \\
&+ \sum_{k=2}^{r-1} x(\tau_k)^T Q_k x(\tau_k) - x(\tau_{k-1})^T Q_k x(\tau_{k-1})
\end{aligned}
$$

Noting $\Psi(t)^T = \left[x(t)^T \; x(\tau_1)^T \; \cdots \; x(\tau_{r-1})^T \right]$ and $\Gamma_i = (A_i^{cl})^T P + P(A_i^{cl})$, the derivative of the Lyapunov–Krasovskii functional candidate becomes:

$$
\begin{aligned}
\dot{V}(x(t)) &= \Psi(t)^T \begin{bmatrix} \sum_{i=1}^{\eta}\gamma_i(t)\Gamma_i + Q_r & 0 & PA_\tau \\ 0 & \bar{Q} & 0 \\ * & 0 & -Q_1 \end{bmatrix} \Psi(t) \\
&= \Psi(t)^T \sum_{i=1}^{\eta}\gamma_i(t) \begin{bmatrix} \Gamma_i + Q_r & 0 & PA_\tau \\ 0 & \bar{Q} & 0 \\ * & 0 & -Q_1 \end{bmatrix} \Psi(t)
\end{aligned}
$$

then from (9.12) it is easy to conclude that $\dot{V}(x(t)) < 0$. Note also that one can write:

$$
\dot{V} < -\min_i \lambda_{min}(\begin{bmatrix} \Gamma_i + Q_r & 0 & PA_\tau \\ 0 & \bar{Q} & 0 \\ * & 0 & -Q_1 \end{bmatrix}) \parallel \Psi \parallel^2 \tag{9.19}
$$

hence, by Krasovskii stability theorem [6], the set $\mathscr{L}_v^p(\rho)$ is given by (9.14) is an invariant set inside the set of attraction of the system. □

Example 9.1 The following example is borrowed from [1] to highlight the importance of partitioning to solve some unfeasible problems. In fact, the obtained conditions with partitioning introduces further degree of freedom that remove some conservativeness of the previous conditions. Consider the delay autonomous system, with a delay $\tau = 1$ given by:

$$
\dot{x}(t) = Ax(t) + A_\tau x(t-\tau)
$$

where

$$
A = \begin{bmatrix} -2.1 & 0 \\ 0 & -0.91 \end{bmatrix}, \quad A_\tau = \begin{bmatrix} -1 & 0 \\ -1 & -1.1 \end{bmatrix}
$$

The LMI of Theorem 9.1 is not feasible. Hence, one can not conclude the stability of the delay system. However, if one uses the partitioning with $r = 2$ by solving the LMI (9.12) of Proposition 9.1, the following solution can be found:

$$X = \begin{bmatrix} 9.2027 & -1.9022 \\ -1.9022 & 9.5772 \end{bmatrix},$$

$$Q_1 = \begin{bmatrix} 14.4942 & -2.1471 \\ -2.1471 & 6.5365 \end{bmatrix} \text{ and } Q_2 = \begin{bmatrix} 28.9205 & -0.7157 \\ -0.7157 & 26.2679 \end{bmatrix}.$$

Finally, one can conclude that the autonomous system with delay is stable.

The previous propositions are useful for stability analysis, one may be also interested in synthesis of such controllers. The synthesis extensions of these results are given as follows:

Proposition 9.2 ([11]) *If there exist matrices $X = X^T \succ 0$, $S = S^T \succ 0$, Y, Z, μ solutions of the following LMIs for $i = 1, ..., \eta$, and $j = 1, ..., m$,*

$$\begin{bmatrix} (AX + B(D_i Y + D_i^- Z))^{sym} + S & A_\tau X \\ * & -S \end{bmatrix} \prec 0 \tag{9.20}$$

$$\begin{bmatrix} \mu & Z_j \\ * & X \end{bmatrix} \succeq 0, \tag{9.21}$$

then the state feedback $F = YX^{-1}$, with matrix $H = ZX^{-1}$ and $\mu = 1/\rho$, is a stabilizing state feedback for the delayed saturating system (9.4) and the set $\mathscr{L}_v(\rho)$ is an invariant set inside the domain of attraction.

Proof The proof will be given in the case of the partitioning below which is more general. □

By using the partitioning the delay interval, the solutions are given by the following result:

Proposition 9.3 ([11]) *If there exist matrices $X = X^T \succ 0$, $S_k = S_k^T \succ 0$, $k = 1, \ldots, r$, Y, Z, and μ solutions of the following LMIs for $i = 1, ..., \eta$, $j = 1, ..., m$,*

$$\begin{bmatrix} (AX + B(D_i Y + D_i^- Z))^{sym} + S_r & 0 & A_\tau X \\ 0 & \bar{S} & 0 \\ * & 0 & -S_1 \end{bmatrix} \prec 0 \tag{9.22}$$

$$\begin{bmatrix} \mu & Z_j \\ * & X \end{bmatrix} \succeq 0, \tag{9.23}$$

where:

$$\bar{S} = \begin{bmatrix} S_1 - S_2 & 0 & \dots & 0 \\ 0 & S_2 - S_3 & & 0 \\ & & \ddots & \\ 0 & 0 & & S_{r-1} - S_r \end{bmatrix},$$

then the state feedback $F = YX^{-1}$, *with matrix* $H = ZX^{-1}$ *and* $\mu = 1/\rho$, *is a stabilizing state feedback for the delayed saturating system (9.4) and the set* $\mathscr{L}_v^P(\rho)$ *is an invariant set inside the domain of attraction.*

Proof The LMI (9.23) guarantees the inclusion $\varepsilon(P, \ \rho) \subset \mathscr{L}(H)$, hence the writing of the saturating system as a convex combination of η delay systems (9.8) is valid. Assume that matrices $X, \ Y, \ Z, \ S_k, k = 1, ..., r$ are the feasible solutions to the LMIs (9.22) and (9.23). Post and pre-multiply the LMI (9.22) by $diag(X^{-1})$ and noting $F = YX^{-1}$, $H = ZX^{-1}$, $P = \rho X^{-1}$, $Q_k = (1/\rho)PS_kP, \ k = 1, ..., r$, leads to the LMI (9.10), for $i = 1, ..., \eta$. Hence, using Proposition 9.1, the solution $x(t) \equiv 0$ is asymptotically stable for all initial condition in the set $\mathscr{L}_v^P(\rho)$. □

9.3.1 α-Stabilizability

As claimed above, the problem of α-stability is also studied. This problem can be viewed in the sense of that the closed-loop time-delay system without saturation is α-stable and the saturating closed-loop system is asymptotically stable for all initial condition in $\mathscr{L}_v(\rho)$.

Let us first define the α-stability:

Definition 9.1 [12] The closed-loop time-delay system (9.4) is α-stable or stable with a rate decay of α if the system

$$\dot{z}(t) = (A + BF + 2\alpha\mathbb{I}_n)z(t) + e^{\alpha\tau}A_\tau z(t-\tau) \tag{9.24}$$

is stable.

Using this definition for the saturating delay systems, the Proposition 9.3 becomes as follows:

Corollary 9.1 *If there exist matrices* $X = X^T \succ 0, \ S_k = S_k^T \succ 0, k = 1, ..., r, \ Y$, Z *and* $\mu = 1/\rho$ *solutions of the following LMIs for* $i = 1, ..., \eta, \quad j = 1, ..., m,$

$$\begin{bmatrix} (AX + B(D_i Y + D_i^- Z))^{sym} + 2\alpha \mathbb{I}_n + S_r & 0 & e^{\alpha\tau} A_\tau X \\ * & \bar{S} & 0 \\ * & 0 & -S_1 \end{bmatrix} \prec 0 \qquad (9.25)$$

$$\begin{bmatrix} \mu & Z_j \\ * & X \end{bmatrix} \succeq 0, \qquad (9.26)$$

where

$$\bar{S} = \begin{bmatrix} S_1 - S_2 & 0 & \dots & 0 \\ 0 & S_2 - S_3 & & 0 \\ & & \ddots & \\ 0 & 0 & & S_{r-1} - S_r \end{bmatrix},$$

then the state feedback $F = YX^{-1}$, with matrix $H = ZX^{-1}$ is a stabilizing state feedback with a rate decay α for the delay saturating system (9.4) for all initial condition in $\mathcal{L}_v^p(\rho)$.

Proof The proof follows the same lines as the proof of Proposition 9.3 applied to the system (9.24). □

Example 9.2 Consider now the closed-loop system given by (9.1), with $\rho = 1, \alpha = 0.3$ and a delay of $\tau = 10$, where:

$$A = \begin{bmatrix} 1 & 1.5 \\ 0.3 & -2 \end{bmatrix}, \quad A_\tau = \begin{bmatrix} 0 & -1 \\ 0 & 0 \end{bmatrix}, \quad B = \begin{bmatrix} 10 \\ 1 \end{bmatrix}$$

$u_{max} = 15 = T$, so matrix B may be changed accordingly with $B^* = BT = [150 \quad 15]^T$ and $u^*(t) = T^{-1}u(t)$. LMIs of Proposition 9.2 ($\mu = 1/\rho$) are not feasible. However, if one uses a partitioning with $r = 9$ and uses LMIs of Proposition 9.3, solution can be found as follows:

$$F = \begin{bmatrix} -0.0107 & -2.5617 \end{bmatrix}^T$$

Figures 9.1 and 9.2, present the state component evolution in time. Whereas Fig. 9.3 shows the control evolution in time.

It can be seen from the examples above, in this first part of the chapter, that partitioning the delay interval in the writing of the Lyapunov–Krasovskii functional leads to a less conservative conditions in both cases of stability or stabilizability of saturating delay systems. Further, to ensure a rate decay for the closed-loop delay system, partitioning may also be easily used to design stabilizing state feedback that guarantees the desired performance of a given rate decay.

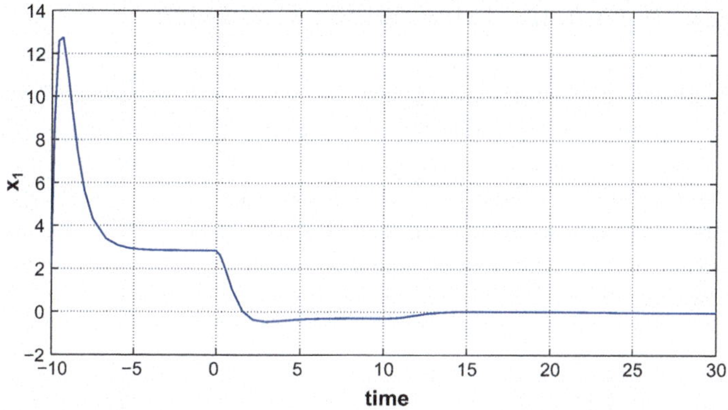

Fig. 9.1 State x_1 evolution

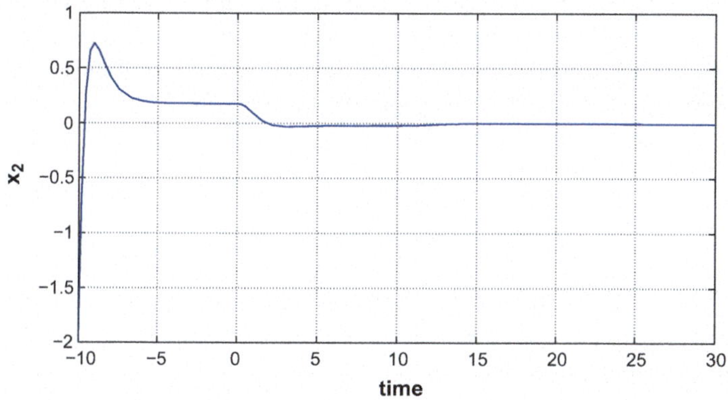

Fig. 9.2 State x_2 evolution

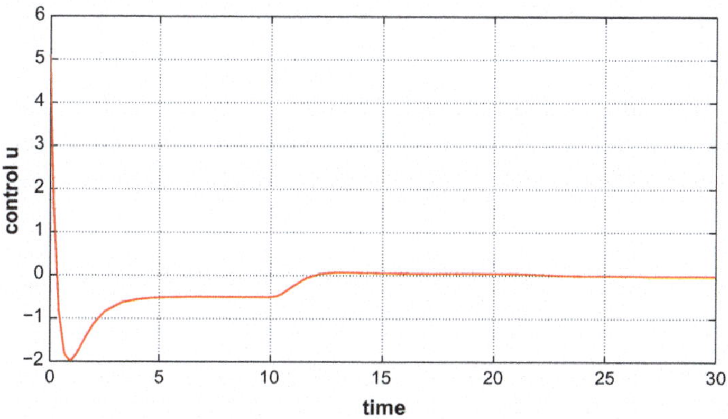

Fig. 9.3 Control evolution

9.4 Delay Dependent Stabilizability Condition

In this second part, the information about the bound on the values that the delay may take can be used. In fact, in some cases, a bound or a maximum value on the delay may be known in advance. In such cases, using an independent delay condition is conservative in the sense that the available information is missed and not used. Using this information leads to delay dependent conditions. In what follows, a delay dependent criterion is worked out to obtain a stabilizing delay dependent condition for the synthesis.

9.4.1 Improved Delay Dependent Condition

An improved delay dependent stability criteria for delay systems presented in [16] is recalled. As pointed out by the authors, this criteria is less conservative than some previous works. Hence, it is worth extending such conditions to obtain a stabilizability condition and also to the case of the saturating system.

Let us consider the autonomous delay system without saturation given by:

$$\begin{cases} \dot{x}(t) &= Ax(t) + A_\tau x(t-\tau) \\ x(t) &= \Phi(t) \text{ for } t \in [-\tau, 0) \end{cases} \tag{9.27}$$

Theorem 9.2 ([16]) *The time-delay system (9.27) is asymptotically stable for any delay τ satisfying $0 < \tau \leq \bar{\tau}$, if there exist matrices $P \succ 0$, $Q \succ 0$, $S \succ 0$, R and W such that the following LMI holds true:*

$$\begin{bmatrix} (PA+R)^{sym} + Q & PA_\tau - R + W^T & -\bar{\tau}R & \bar{\tau}A^T S \\ * & -Q - W - W^T & -\bar{\tau}W & \bar{\tau}A_\tau^T S \\ * & * & -\bar{\tau}S & 0 \\ * & * & * & -\bar{\tau}S \end{bmatrix} \prec 0 \tag{9.28}$$

As claimed, the delay dependent stability criteria presented above is only useful for analysis. In what follows, it is extended to controlled systems to enable the synthesis of such stabilizing controllers. Consider the controlled delay system without saturation

$$\begin{cases} \dot{x}(t) &= Ax(t) + A_\tau x(t-\tau) + Bu(t) \\ x(t) &= \Phi(t) \text{ for } t \in [-\tau, 0) \end{cases} \tag{9.29}$$

Proposition 9.4 *If there exist matrices $X \succ 0$, $\bar{Q} \succ 0$, $\Sigma \succ 0$, $\bar{R}$ and $\bar{W}$ such that the LMI below holds true, then the time-delay system (9.29) is asymptotically stable for any delay τ satisfying $0 < \tau \leq \bar{\tau}$,*

$$\begin{bmatrix} (AX + BY + \bar{R})^{sym} + \bar{Q} & A_\tau X - \bar{R} + \bar{W}^T & -\bar{\tau}\bar{R} & * \\ * & -\bar{Q} - \bar{W} - \bar{W}^T & -\bar{\tau}\bar{W} & \bar{\tau}XA_\tau^T \\ * & * & -2\bar{\tau}X + \bar{\tau}\Sigma & 0 \\ \bar{\tau}AX + \bar{\tau}BY & * & * & -\bar{\tau}\Sigma \end{bmatrix} \prec 0 \quad (9.30)$$

Furthermore, the controller is given by $F = YX^{-1}$.

Proof Using a memoryless state feedback $u(t) = Fx(t)$ and replacing A by $A + BF$ in (9.27), the closed-loop system becomes an autonomous system. Furthermore, the asymptotic stability condition is true if one replaces A by $A + BF$ in (9.28). Noting $X = P^{-1}$, $\bar{R} = XRX$, $\bar{W} = XWX$, $\bar{Q} = XQX$, $\Sigma = S^{-1}$ and post and pre-multiplying by $diag\{X, X, X, S^{-1}\}$ leads to the following LMI:

$$\begin{bmatrix} (AX + BY + \bar{R})^{sym} + \bar{Q} & A_\tau X - \bar{R} + \bar{W}^T & -\bar{\tau}\bar{R} & * \\ * & -\bar{Q} - \bar{W} - \bar{W}^T & -\bar{\tau}\bar{W} & \bar{\tau}XA_\tau^T \\ * & * & -\bar{\tau}XSX & 0 \\ \bar{\tau}AX + \bar{\tau}BY & * & * & -\bar{\tau}\Sigma \end{bmatrix} \prec 0$$

bearing in mind that the following inequality is true

$$-XSX \preceq -2X + S^{-1}$$

leads to the LMI (9.30). □

Using the convexity property one may write (9.8) as follows:

$$\begin{cases} \dot{x}(t) = \sum_{i=1}^{\eta} \gamma_i(t)(A + B(D_i F + D_i^- H))x(t) + A_\tau x(t-\tau) \\ x(t) = \Phi(t) \text{ for } t \in [-\tau, 0) \end{cases} \quad (9.31)$$

9.4.2 Delay Dependent Condition for Saturating Systems

The stability analysis by the criteria given is firstly extended to controlled systems and a stabilizability condition is derived. Hence, the writing of the saturated delay system as a convex combination of some delay systems enables the improved stabilizability condition established below to be extended to the case of saturating delay systems. These conditions are obtained in LMI form and can be easily solved using the existing toolboxes software of the LMI environment.

Taking into account the saturating term at the input of the delay system, the following proposition may be given:

Proposition 9.5 *If there exist matrices* $X \succ 0$, $\bar{Q} \succ 0$, $\Sigma \succ 0$, $\bar{R}$ *and* $\bar{W}$ *such that for* $i = 1, \ldots, \eta$, *and* $j = 1, \ldots, m$

$$\begin{bmatrix} \Pi_i & A_\tau X - \bar{R} + \bar{W}^T & -\bar{\tau}\bar{R} & \bar{\tau}XA^T + \bar{\tau}Y^T D_i B^T + \bar{\tau}Z^T D_i^- B^T \\ * & -\bar{Q} - \bar{W} - \bar{W}^T & -\bar{\tau}\bar{W} & \bar{\tau}XA_\tau^T \\ * & * & -2\bar{\tau}X + \bar{\tau}\Sigma & 0 \\ * & * & * & -\bar{\tau}\Sigma \end{bmatrix} \prec 0 \quad (9.32)$$

$$\begin{bmatrix} 1 & Z_j \\ * & X \end{bmatrix} \succeq 0, \quad (9.33)$$

where

$$\Pi_i = (AX + \bar{R} + BD_iY + BD_i^- Z)^{sym} + \bar{Q},$$

then the time-delay system (9.29) is delay dependent asymptotically stable for any delay τ satisfying $0 < \tau \leq \bar{\tau}$. Further the controller is given by $F = YX^{-1}$. Furthermore, the set $\mathcal{L}_v(\rho)$ is invariant inside the set of attraction of the system.

Proof First, let us note $Z = HX^{-1}$ and consider as in [16], the Lyapunov–Krasovskii functional given by:

$$\begin{aligned} V(x_t) &= V_1(x(t)) + V_2(x(t)) + V_3(x(t)) \quad (9.34) \\ V_1(x_t) &= x(t)^T P x(t), \\ V_2(x_t) &= \int_{-\tau}^{0}\int_{t+\beta}^{t} \dot{x}(\sigma)^T S \dot{x}(\sigma) d\sigma d\beta \\ V_3(x_t) &= \int_{t-\tau}^{t} x(\sigma)^T Q x(\sigma) d\sigma. \end{aligned}$$

Define $\mathcal{L}_v(\rho)$ as:

$$\mathcal{L}_v(\rho) = \{\ \phi \in \mathcal{C}_{n,\tau} /\ \ V(\phi(t)) \leq \rho\}. \quad (9.35)$$

Along same lines as in [16], it follows that:

$$\dot{V}(x_t) = \frac{1}{\tau}\sum_{i=1}^{\eta} \gamma_i(t) \int_{t-\tau}^{t} \xi(t,\sigma)^T \Lambda_i \xi(t,\sigma) d\sigma \quad (9.36)$$

where $\xi(t,\sigma) = [x(t)^T \ \ x(t-\tau)^T \ \ \dot{x}(\sigma)^T]^T$ and

$$\Lambda_i = \begin{bmatrix} \Gamma_i & A_\tau X - \bar{R} + \bar{W}^T & -\bar{\tau}\bar{R} & \bar{\tau}XA^T + \bar{\tau}Y^T D_i B^T + \bar{\tau}Z^T D_i^- B^T \\ * & -\bar{Q} - \bar{W} - \bar{W}^T & -\bar{\tau}\bar{W} & \bar{\tau}XA_\tau^T \\ * & * & -2\bar{\tau}X + \bar{\tau}S & 0 \\ * & * & * & -\bar{\tau}S \end{bmatrix} \quad (9.37)$$

where:

$$\Gamma_i = (AX + BD_iY + BD_i^- Z + \bar{R})^{sym} + \bar{\tau} A^T SA + \bar{Q}.$$

Using the Lyapunov–Krasovskii Theorem [9], for all $x_t \in \mathscr{L}_v(\rho)$, one has to prove that there exist continuous positive non decreasing scalar functions ε_1, ε_2, and ε_3 such that:

$$\begin{aligned} \varepsilon_1(\|x_t(0)\| \leq V(x_t) \leq \varepsilon_2(\|x_t\|_c) \\ \dot{V}(x_t) \leq -\varepsilon_3(\|x_t(0)\|); \end{aligned}$$

for the proposed function $V(x_t)$ given by (9.34), one can write

$$\varepsilon_1 \|x_t(0)\|^2 \leq V(x_t) \leq \varepsilon_2 \|x_t\|_c^2$$

where:

$$\begin{aligned} \varepsilon_1 &= \lambda_{min}(P) \\ \varepsilon_2 &= \lambda_{max}(P) + \tau\lambda_{max}(Q) + (\tau^2/2)(\max_{i=1,\ldots\eta} \left\| A + BD_iY + BD_i^- Z \right\| + \|A_\tau\|)\lambda_{max}(S). \end{aligned}$$

On the other hand, from (9.36), it is possible to write that:

$$\dot{V}(x_t) \leq (-\min_{i=1,\ldots,\eta} (\lambda_{min}(-\Lambda_i)))\, \|x_t\|^2 . \tag{9.38}$$

□

Example 9.3 • nonsaturating input case:
Consider the nonsaturating input delay system given by

$$A = \begin{bmatrix} 0 & 0 \\ 0 & 1 \end{bmatrix}; \quad A_\tau = \begin{bmatrix} -1 & -1 \\ 0 & -0.9 \end{bmatrix}; \quad B = \begin{bmatrix} 0 \\ 1 \end{bmatrix},$$

when considering the nonsaturating case, applying Proposition 9.4 leads to a less conservative bound delay $\bar{\tau} = 1.7$ compared to some previous works. Delay bounds together with the stabilizing state feedback are summarized in Table 9.1. The solution of LMI (9.30) for nonsaturating input delay system gives the following results:

Table 9.1 Comparative table

Method	$\bar{\tau}$	F
[5]	1.4	not given
[4]	1.5	[−58.3 − 294.9]
Proposition 9.4	1.7	[−0.2991 − 2.1502]

$$X = \begin{bmatrix} 518.2308 & -126.6460 \\ -126.6460 & 40.2598 \end{bmatrix}, \quad \Sigma = \begin{bmatrix} 346.6682 & -84.6458 \\ -84.6458 & 39.1424 \end{bmatrix},$$
$$\bar{Q} = \begin{bmatrix} 390.0560 & -95.2633 \\ -95.2633 & 34.1351 \end{bmatrix}, \quad \bar{R} = \begin{bmatrix} -402.5896 & 98.4295 \\ 98.4029 & -24.1430 \end{bmatrix},$$
$$\bar{W} = \begin{bmatrix} 403.0770 & -98.5495 \\ -98.5460 & 24.1784 \end{bmatrix},$$
$$Y = \begin{bmatrix} 117.3107 & -48.6869 \end{bmatrix} \; and \; F = \begin{bmatrix} -0.2991 & -2.1502 \end{bmatrix}.$$

Figures 9.4 and 9.5 present, respectively, state and control evolution for non-saturating input system with initial conditions $x_o = [-5 \quad 5]^T$ and delay $\tau = 1.65s$.

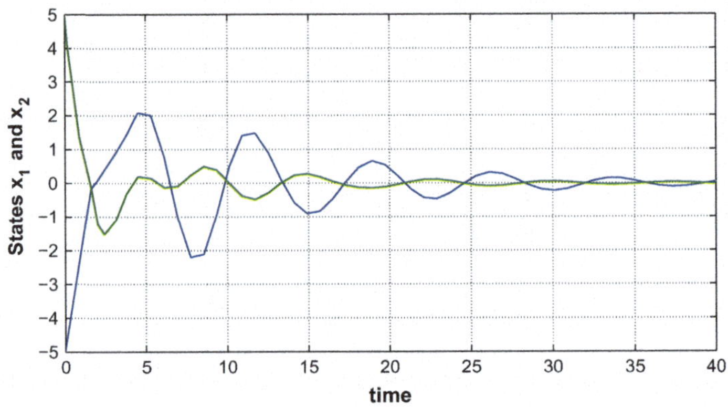

Fig. 9.4 States x_1 and x_2 evolution

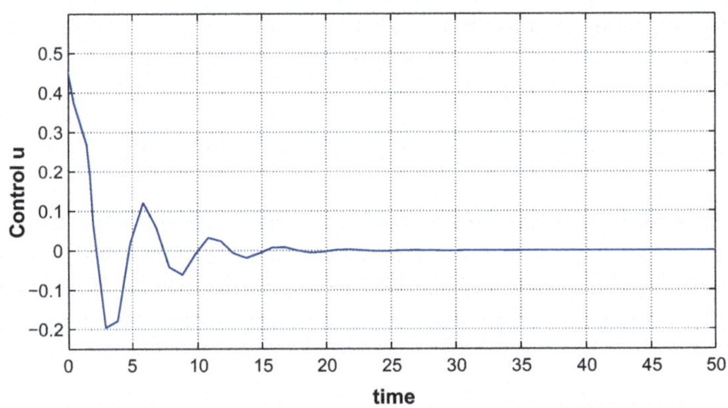

Fig. 9.5 Control evolution

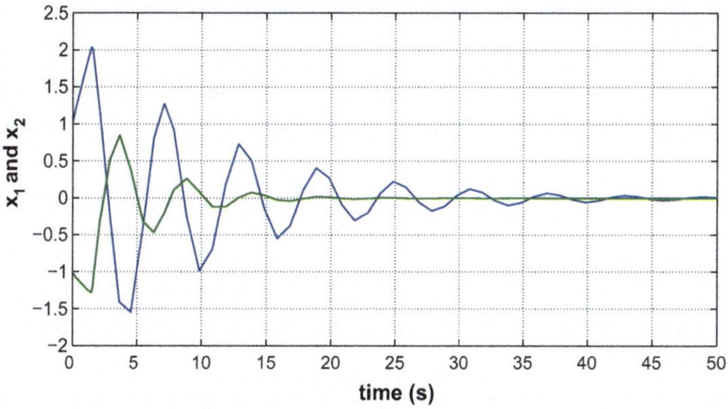

Fig. 9.6 States evolution x_1 and x_2

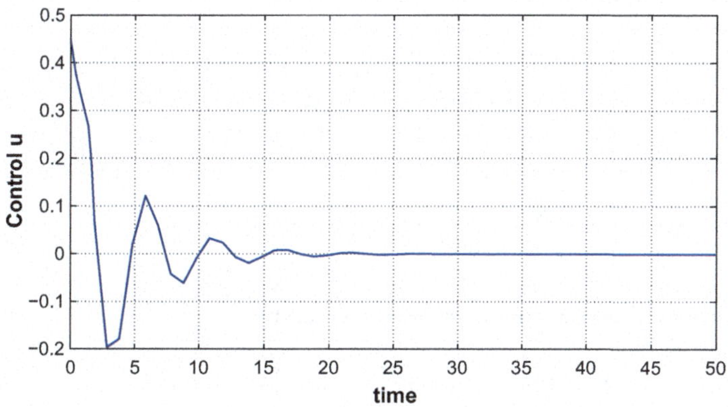

Fig. 9.7 Control evolution

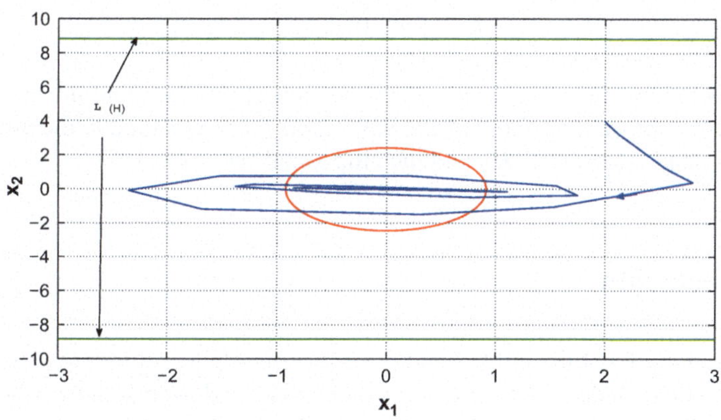

Fig. 9.8 Trajectory of the system in the state space

• Saturating input case:

Let us now consider the saturating input delay system given by (9.31) taken from [13] where we consider no perturbation and a control bound $u_{max} = 15$, where

$$A = \begin{bmatrix} -0.2 & 0 \\ 0 & 1 \end{bmatrix}; \quad A_\tau = \begin{bmatrix} -1 & 0 \\ 0 & -1 \end{bmatrix}; \quad B = \begin{bmatrix} 0 \\ 1 \end{bmatrix}.$$

The non unity of the control limitation is easily accommodated by taking:

$$B_n = \begin{bmatrix} 0 \\ 15 \end{bmatrix}$$

For delays satisfying $0 < \tau < 1.57s$, that is $\bar{\tau} = 1.57s$, LMIs (9.32) and (9.33) are found feasible, and this leads to the following data:

$$X = \begin{bmatrix} 4.9800 & 0 \\ 0 & 5.9013 \end{bmatrix}, \quad \bar{Q} = \begin{bmatrix} 4.9247 & 0 \\ 0 & 5.6226 \end{bmatrix},$$
$$\bar{S} = \begin{bmatrix} 4.7486 & 0 \\ 0 & 8.8124 \end{bmatrix}, \quad \bar{R} = \begin{bmatrix} -3.3277 & 0 \\ 0 & -1.5515 \end{bmatrix};$$
$$\bar{W} = \begin{bmatrix} 3.2783 & 0 \\ 0 & 1.5985 \end{bmatrix};$$

the two matrices Y and Z are, respectively, given by:

$$Y = \begin{bmatrix} 0 & -0.6689 \end{bmatrix}, \quad Z = \begin{bmatrix} 0 & -0.6682 \end{bmatrix}$$

hence the stabilizing memoryless state feedback is given by:

$$F = \begin{bmatrix} 0 & -0.1133 \end{bmatrix}.$$

The figures below show the evolution of the saturating input and the states versus time, respectively, on Figs. 9.6, 9.7 and 9.8. Figure 9.8 is reserved to the trajectory of the system inside the set $\mathscr{L}(H)$ and the ellipsoidal set $\varepsilon(X, 1)$.

9.5 Conclusion

The elaboration of stabilizability conditions for time-delay saturating actuators systems are studied. Hence, LMIs are derived enabling to design stabilizing state feedback for such systems. First, partition of the delay interval is introduced. This gives more degree of freedom to overcome some conservativeness of previous conditions in former works in delay independent cases. In fact, partitioning the time-delay interval for the Lyapunov functional renders some unfeasible problems feasible as stated in [1] and as presented here for two examples. Further, the issue of introducing a rate decay for stability, in closed-loop, is easily deduced and also written into LMI form. On the other hand, and as a second part, a delay dependent stability criteria is worked out to derive delay dependent stabilizability conditions for autonomous systems without and with saturating input. Further, the case of the non-saturating delay systems is compared to some previous results. The conditions enabling synthesis of stabilizing state feedback for such systems are given under LMI formalism. Each part is concluded with illustrative examples that show the effectiveness of the proposed methods. A parametric Lyapunov equation approach can also be used to study stabilization of linear systems with input delay and saturation [17]. Other extensions exist to solve the problem of stabilization of linear systems with distributed input delay and input saturation [19].

References

1. E.K. Boukas, Z.K. Liu, Deterministic and stochastic time-delay systems, in *Control Engineering* (Birkhauser, 2002)
2. Y.Y. Cao, Z. Lin, T. Hu, Stability analysis of linear time delay systems subject to input saturation. IEEE Trans. Circuit Syst. I Fundam. Theory Appl. **49**, 233–240 (2002)
3. B.S. Chen, S.S. Wang, H.C. Lu, Stabilization of time delay systems containing saturating actuators. Int. J. Control **47**, 867–881 (1988)
4. E. Fridman, U. Shaked, An improved stabilization method for linear time delay systems. IEEE Trans. Autom. Control **47**(11), 1931–1937, (2002)
5. E. Fridman, U. Shaked, Delay dependent stability and H_∞ control: constant and time varying delays. Int. J. Control **76**, 48–90 (2003)
6. J. Hale, *Theory of Functional Differential Equations* (Springer, New York, 1977)
7. A. Hmamed, A. Benzaouia, H. Bensaleh, Regulator problem for linear continuous-time delay systems with nonsymmerical constrained control. IEEE Trans. Autom. Control **40**, 1615–1619 (1995)
8. T. Hu, Z. Lin, Control Systems with actuator saturation, in *Control Engineering* (Birkhauser, 2001)
9. N.N. Krasovskii, *Stability of Motion.* (Stanford University Press, CA, 1963)
10. F. Mesquine, A. Benzaouia, M. Benhayoun, F. Tadeo, Further results on stabilizability for time delay systems with saturating control, in *11th International Conference on Sciences and Techniques of Automatic Control and Computer Engineering*, Monastir, Tunisia (2010)
11. F. Mesquine, M. Benhayoun, A. Benzaouia, F. Tadeo, Improved stability condition for time delay systems with saturating control, in *2nd International Conference on Sciences and Control*, Marrakech, Morocco (2012)
12. T. Mori, N. Fukuma, M. Kuwahara, On an estimate of the decay rate for stable linear delay systems. Int. J. Control **36**, 95–97 (1982)
13. S. Oucheriah, Synthesis of controllers for time delay systems subject to actuator saturation and disturbance. J. Dyn. Syst. Meas. Control **125**, 244– 249 (2003)
14. S. Oucheriah, Robust exponential convergence of a class of linear delayed systems with bounded controllers and disturbances. Automatica **42**, 1863–1867 (2006)
15. E. Tissir, A. Hmamed, Further results on the stabilization of time delay systems containing saturating actuators. Int. J. Syst. Sci. **23**, 615–622 (1992)
16. S. Xu, J. Lam, Improved delay dependent stability criteria for time delay systems. IEEE Trans. Autom. Control **50**(3), 384–387 (2005)

17. B. Zhou, Z. Lin, G. Duan, Stabilization of linear systems with input delay and saturation-A parametric Lyapunov equation approach. Int. J. Robust Nonlinear Control **20**(13), 1502–1519 (2009)
18. B. Zhou, Z. Lin, Parametric Lyapunov equation approach to stabilization of discrete-time systems with input delay and saturation. IEEE Trans. Circuits Syst. 1 Regul. Pap. **58-1**(11), 2741–2754 (2011)
19. B. Zhou, H. Gao, Z. Lin, G.R. Duan, Stabilization of linear systems with disturbed input delay and input saturation. Automatica **48**(5), 712–724 (2012)

Chapter 10
Stabilization of 2D Continuous Systems with Multi-delays and Saturated Control

10.1 Introduction

This chapter studies the stability of linear two-dimensional (2D) continuous systems with multi-delays and input saturation. In the last two decades, the 2D system theory has been paid considerable attention by many researchers [13]. The 2D linear models were introduced in the seventies [7, 8] and have found many applications in digital data filtering, image processing [20], modeling of partial differential equations [15], etc. In connection with the Roesser models [20] and the Fornasini-Marchesini models [6], some important problems, such as realization, control, minimum energy control, have been extensively investigated (see, e.g., [14]). The stabilization problem is not fully investigated and still not completely solved [3].

This chapter is interested to saturated continuous 2D systems with multi-delays described by Roesser model. This extension to systems with delay is prompted by the existence of transport delays in many problems in process control. The stabilization problem is of interest, because the existence of a delay might cause instabilities [16]. The stabilization of this kind of systems has been extensively studied in the literature for 1D (see [10, 11, 16] and the references therein). This problem has already been studied for 2D systems by considering independent and dependent stability and stabilization conditions [17, 18]. However, all the studies on 2D delay systems are only available for discrete systems, except authors in [5], where the 2D continuous systems with delay is taken without saturation. To the best of our knowledge, no works on saturated 2D continuous systems with delay exist before the work of [2, 4].

The objective of this work is the design of stabilizing state feedback controllers for this class of systems. To the best of the authors knowledge, no works have directly considered saturated 2D continuous systems with delay. To this end, quadratic Lyapunov functions are used. In this context, sufficient conditions of stabilizability under LMI form are presented. This formulation enables one to derive saturated state feedback controllers.

A. Benzaouia et al., *Saturated Control of Linear Systems*, Studies in Systems, Decision and Control 124, https://doi.org/10.1007/978-3-319-65990-9_10

10.2 Problem Formulation

Consider the 2D continuous system described by the Roesser model with multi-delays and saturated control:

$$\begin{bmatrix} \frac{\partial x^h(t_1,t_2)}{\partial t_1} \\ \frac{\partial x^v(t_1,t_2)}{\partial t_2} \end{bmatrix} = Ax(t_1,t_2) + \sum_{i=1}^{r} G_i x(t_1 - \tau_i^h, t_2 - \tau_i^v) + Bsat(u(t_1,t_2)) \tag{10.1}$$

$$\begin{aligned} x^h(0,t_2) &= f(t_2),\, t_2 \in \left[-\tau_{max}^v, 0\right],\, \tau_{max}^v = max(\tau_i^v) \\ x^v(t_1,0) &= g(t_1),\, t_1 \in \left[-\tau_{max}^h, 0\right],\, \tau_{max}^h = max(\tau_i^h) \\ x_o &= \begin{bmatrix} f(t_2) \\ g(t_1) \end{bmatrix}, \end{aligned}$$

with

$$x(t_1,t_2) \triangleq \begin{bmatrix} x^h(t_1,t_2) \\ x^v(t_1,t_2) \end{bmatrix}, \tag{10.2}$$

$$x(t_1 - \tau_i^h, t_2 - \tau_i^v) \triangleq \begin{bmatrix} x^h(t_1 - \tau_i^h, t_2) \\ x^v(t_1, t_2 - d_i) \end{bmatrix},$$

$$A = \begin{bmatrix} A_{11} & A_{12} \\ A_{21} & A_{22} \end{bmatrix}, G_i = \begin{bmatrix} G_{i11} & G_{i12} \\ G_{i21} & G_{i22} \end{bmatrix}, B = \begin{bmatrix} B_1 \\ B_2 \end{bmatrix}$$

where $x^h(t_1,t_2)$ is the horizontal state in $\mathbb{R}^{n_h}$, $x^v(t_1,t_2)$ is the vertical state in $\mathbb{R}^{n_v}$, $u(t_1,t_2)$ is the control vector in $\mathbb{R}^m$ and $g(t_1), f(t_2)$ present the boundary conditions, and τ_i^h and τ_i^v are the delays in horizontal and vertical directions, respectively. Note $n = n_h + n_v$. r in Eq. (10.1) represents the number of delays in any direction.

It is assumed that the 2D system is stabilizable. Consider the standard saturation defined as follows: for $i = 1, \ldots, m$;

$$sat(u) = (sat(u_i)) = \begin{cases} 1 & \text{if } u_i > 1 \\ u_i & \text{if } -1 \le u_i \le 1 \\ -1 & \text{if } u_i < -1 \end{cases} \tag{10.3}$$

Further, consider the state feedback control:

$$u(t_1,t_2) = Kx(t_1,t_2) \tag{10.4}$$

where matrix $K = [K_1 \quad K_2]$ is the state feedback gain we are looking for. The problem is to compute a static feedback control given by (10.4) such that the saturated closed-loop 2D system with multiple delays in (10.1) is asymptotically stable. Furthermore, define the sets $\varepsilon(P, \ \rho)$ and $\mathscr{L}(H)$ as follows:

$$\varepsilon(P,\ \rho) = \{x \in \mathbb{R}^n / x^T P x \le \rho;\quad P = P^T \succ 0\} \tag{10.5}$$

$$\mathscr{L}(H) = \{x \in \mathbb{R}^n / |H_l x| \le 1\ ,\ l = 1, \ldots, m\} \tag{10.6}$$

where H_l denotes the lth row of matrix H, and ρ is a positive scalar. The problem we address thereafter is to find stabilizing state feedback controllers for the 2D continuous systems (10.1) with saturation on the control and multi-delays. First, conditions of stabilizability are established. Second, these conditions are worked out to give a procedure enabling the stabilizing controller (10.4) to be computed.

10.3 Some 2D Extensions

This section is devoted to some preliminaries useful to the development below: The first lemma enables the saturated 2D system with multi-delays to be written as a convex combination of $\eta = 2^m$ linear systems. Conditions of stability for 2D linear systems are then presented. Finally, a technical lemma providing a sufficient condition of stability is given.

The result of Lemma 1.3 can be extended to 2D systems since the reasoning depends on the saturation function and not on the number of dimensions (independent variables on which the control depends). Thus, for 2D systems, the saturation function can be written as follows:

$$sat(u(t_1, t_2)) = \sum_{s=1}^{\eta} \gamma_s (D_s u(t_1, t_2) + D_s^- v(t_1, t_2)) \tag{10.7}$$

Hence, in state feedback control, for two given feedback matrices K and H with $u = Kx$ and $v = Hx$, such that $|H_l x| < 1$. Then, one can write:

$$sat(Kx(t_1, t_2)) \ = \ \sum_{s=1}^{\eta} \gamma_s (D_s K x(t_1, t_2) + D_s^- H x(t_1, t_2)) \tag{10.8}$$

Consider now the following 2D continuous autonomous system:

$$\begin{bmatrix} \frac{\partial x^h(t_1,t_2)}{\partial t_1} \\ \frac{\partial x^v(t_1,t_2)}{\partial t_2} \end{bmatrix} = \bar{A} x(t_1, t_2). \tag{10.9}$$

Theorem 10.1 [9, 19] *The 2D linear continuous system (10.9) is asymptotically stable if there exists a positive definite matrix P of the form:*

$$P = \begin{bmatrix} P_1 & 0 \\ 0 & P_2 \end{bmatrix}, \tag{10.10}$$

with $P_1 \in \mathbb{R}^{n_h \times n_h}$ *and* $P_2 \in \mathbb{R}^{n_v \times n_v}$, *such that:*

$$\bar{A}^T P + P\bar{A} \prec 0 \tag{10.11}$$

In this case, the following function

$$\begin{aligned} V(x(t_1,t_2)) &= x^T(t_1,t_2)Px(t_1,t_2) \\ &= x^{hT}(t_1,t_2)P_1x^h(t_1,t_2) + x^{vT}(t_1,t_2)P_2x^v(t_1,t_2) \\ &= V_1(t_1,t_2) + V_2(t_1,t_2) \end{aligned} \tag{10.12}$$

is a Lyapunov function of the system.

This result can also be proved by using function (10.12) together with the following definition presented for the first time in [12].

Definition 10.1 The derivative of function $V(t_1, t_2)$ given by:

$$\dot{V}_u(t_1,t_2) = \frac{\partial V_1(t_1,t_2)}{\partial t_1} + \frac{\partial V_2(t_1,t_2)}{\partial t_2} \tag{10.13}$$

is called the unidirectional derivative.

Note that this unidirectional derivative can be seen as a particular case of the derivative of the function $V(t_1, t_2)$ in one direction (respectively, t_1 or t_2), independently of the other (respectively, t_2 or t_1).

Lemma 10.1 [12] *The 2D continuous linear system (10.9) is asymptotically stable if its unidirectional derivative (10.13) is negative.*

Proof The idea of the proof is based on the negativity of (10.13) which implies condition (10.11). □

10.4 2D-Constrained Control with Delays

Reference [1] with the background of the previous sections, sufficient conditions are now given for the stabilization of saturated 2D continuous systems with multi-delays. The problem is to compute a static feedback control given by (10.4) such that the closed-loop 2D system with multi-delays is asymptotically stable.

Using the state feedback control (10.8) and the fact that $v = Hx$ with $x \in \mathcal{L}(H)$, the 2D saturated continuous system with multi-delays, described by the Roesser model, is presented in the next equation:

$$\begin{bmatrix} \frac{\partial x^h(t_1,t_2)}{\partial t_1} \\ \frac{\partial x^v(t_1,t_2)}{\partial t_2} \end{bmatrix} = Ax(t_1, t_2) + \sum_{i=1}^{r} G_i x(t_1 - \tau_i^h, t_2 - \tau_i^v)$$

$$+B \sum_{s=1}^{\eta} \gamma_s(t_1, t_2)(D_s K + D_s^- H)x(t_1, t_2).$$

That is,

$$\begin{bmatrix} \frac{\partial x^h(t_1,t_2)}{\partial t_1} \\ \frac{\partial x^v(t_1,t_2)}{\partial t_2} \end{bmatrix} = \sum_{s=1}^{\eta} \gamma_s(t_1, t_2)\tilde{A}_s x(t_1, t_2) + \sum_{i=1}^{r} G_i x(t_1 - \tau_i^h, t_2 - \tau_i^v) \quad (10.14)$$

$$= \tilde{A}(\gamma)x(t_1, t_2) + \sum_{i=1}^{r} G_i x(t_1 - \tau_i^h, t_2 - \tau_i^v)$$

where matrix $\tilde{A}_s$ and $\tilde{A}(\gamma)$ are given by:

$$\tilde{A}_s = \begin{bmatrix} \tilde{A}_{s11} & \tilde{A}_{s12} \\ \tilde{A}_{s21} & \tilde{A}_{s22} \end{bmatrix},$$

$$\tilde{A}(\gamma) = \sum_{s=1}^{\eta} \gamma_s(t_1, t_2)\tilde{A}_s. \quad (10.15)$$

with: $\tilde{A}_{sij} = A_{ij} + B_i(D_s K_j + D_s^- H_j),\ i, j = 1 \text{ or } 2$ and $s = 1, \ldots, \eta$.

The sufficient conditions for system (10.14) to be stabilizable are stated by the following result:

Theorem 10.2 *If there exist symmetric matrices* $P_1 \succ 0$, $P_2 \succ 0$, $Q_1 \succ 0$, ..., $Q_\eta \succ 0$, $R_1 \succ 0$, ..., $R_\eta \succ 0$, *and matrices* K_1, K_2, H_1, H_2 *and scalar* $\rho > 0$ *such that the following conditions:*

$$\varepsilon(P, \ \rho) \subset \mathscr{L}(H), \quad (10.16)$$

$$\begin{bmatrix} (\Xi)_{11} & (\Xi)_{12} & P_1(G_1)_{11} & P_1(G_1)_{12} & P_1(G_r)_{11} & P_1(G_r)_{12} \\ * & (\Xi)_{22} & P_2(G_1)_{21} & P_2(G_1)_{22} & P_2(G_r)_{21} & P_2(G_r)_{22} \\ * & * & -Q_1 & 0 & 0 & 0 \\ * & * & 0 & -R_1 & 0 & 0 \\ .. & .. & .. & .. & .. & .. \\ * & * & * & * & -Q_r & 0 \\ * & * & * & * & * & -R_r \end{bmatrix} \prec 0 \quad (10.17)$$

$$\forall s \in [1, \eta]$$

hold true, then 2D saturated continuous system with multi-delays (10.14) is asymptotically stable,
with:

$$(\Xi)_{11} = [P_1A_{11} + P_1B_1(D_sK_1 + D_s^-H_1)]^{sym} + \sum_{i=1}^{\eta} Q_i$$

$$(\Xi)_{12} = P_1A_{12} + A_{21}^TP_2 + P_1B_1(D_sK_2 + D_s^-H_2) + (D_sK_1 + D_s^-H_1)^TB_2^TP_2$$

$$(\Xi)_{22} = [P_2A_{22} + P_2B_2(D_sK_2 + D_s^-H_2)]^{sym} + \sum_{i=1}^{\eta} R_i$$

Proof Assume that condition (10.16) holds. In this case, $|H_lx(t_1, t_2)| \leq 1, l = 1, \ldots, m$. This allows the saturation to be written according to (10.7) using (10.8). Thus, the closed-loop system is given by (10.14). Consider now the following Lyapunov–Krasovskii functional:

$$\begin{aligned} V_1(t_1, t_2) &= x^{hT}(t_1, t_2)P_1x^h(t_1, t_2) + \sum_{i=1}^{r} \int_{t_1-\tau_i^h}^{t_1} x^{hT}(\sigma, t_2)Q_ix^h(\sigma, t_2)d\sigma \\ V_2(t_1, t_2) &= x^{vT}(t_1, t_2)P_2x^v(t_1, t_2) + \sum_{i=1}^{r} \int_{t_2-\tau_i^v}^{t_2} x^{vT}(t_1, \sigma)R_ix^v(t_1, \sigma)d\sigma \end{aligned} \tag{10.18}$$

For the system with multiple delays in closed-loop, the Lyapunov–Krasovskii candidate functional is:

$$V(t_1, t_2) = V_1(t_1, t_2) + V_2(t_1, t_2)$$

Computing directional derivatives of V_1 and V_2 gives:

$$\begin{aligned} \frac{\partial V_1(t_1, t_2)}{\partial t_1} &= \frac{\partial x^{hT}}{\partial t_1}P_1x^h + x^{hT}P_1\frac{\partial x^h}{\partial t_1} + x^{hT}(\sum_{i=1}^{r} Q_i)x^h \\ &\quad - \sum_{i=1}^{r} x^{hT}(t_1 - \tau_i^h, t_2)Q_ix^h(t_1 - \tau_i^h, t_2) \end{aligned} \tag{10.19}$$

$$\begin{aligned} \frac{\partial V_2(t_1, t_2)}{\partial t_2} &= \frac{\partial x^{vT}}{\partial t_2}P_2x^v + x^{vT}P_2\frac{\partial x^v}{\partial t_2} + x^{vT}(\sum_{i=1}^{r} R_i)x^v \\ &\quad - \sum_{i=1}^{r} x^{vT}(t_1, t_2 - \tau_i^v)R_ix^v(t_1, t_2 - \tau_i^v) \end{aligned} \tag{10.20}$$

Summing up (10.19)–(10.20) enables the unidirectional derivative of the Lyapunov–Krasovskii functional to be wtitten as:

$$\dot{V}_u(t_1,t_2) = x^T(\tilde{A}^T(\gamma)P + P\tilde{A}(\gamma))x + x^T(\sum_{i=1}^{\eta}\Gamma_i)x + (\sum_{i=1}^{r} x^T(t_1-\tau_i^h, t_2-\tau_i^v)G_i^T)Px$$
$$+ x^T P(\sum_{i=1}^{r} G_i x(t_1-\tau_i^h, t_2-\tau_i^v)) - \sum_{i=1}^{r} x^T(t_1-\tau_i^h, t_2-\tau_i^v)\Gamma_i x(t_1-\tau_i^h, t_2-\tau_i^v) \quad (10.21)$$

with:

$$\Gamma_i = \begin{bmatrix} Q_i & 0 \\ 0 & R_i \end{bmatrix},$$

Substituting (10.15) in (10.21) and bearing in mind that $\sum_{s=1}^{\eta}\gamma_s = 1$, $\gamma_s \geqslant 0$, $s = 1, \ldots, \eta$ lead to:

$$\dot{V}_u(t_1,t_2) = x^T(\sum_{s=1}^{\eta}\gamma_s(\tilde{A_s}^T P + P\tilde{A_s}))x + x^T(\sum_{i=1}^{\eta}\Gamma_i)x + (\sum_{i=1}^{r} x^T(t_1-\tau_i^h, t_2-\tau_i^v)G_i^T)Px$$
$$+ x^T P(\sum_{i=1}^{r} G_i x(t_1-\tau_i^h, t_2-\tau_i^v)) - \sum_{i=1}^{r} x^T(t_1-\tau_i^h, t_2-\tau_i^v)\Gamma_i x(t_1-\tau_i^h, t_2-\tau_i^v) \quad (10.22)$$

Introduce the augmented state ξ as:

$$\xi(t_1,t_2) = \begin{bmatrix} x(t_1,t_2) \\ x(t_1-\tau_1^h, t_2-\tau_1^v) \\ \cdots \\ x(t_1-\tau_r^h, t_2-\tau_r^v) \end{bmatrix}$$

In this case, the unidirectional derivative of Lyapunov–Krasovskii functional can be written as:

$$\dot{V}_u(t_1,t_2) = \sum_{s=1}^{\eta}\gamma_s\xi^T(t_1,t_2)M_s\xi(t_1,t_2), \quad (10.23)$$

with M_s defined by:

$$M_s = \begin{bmatrix} M_0(s) & PG_1 & . & . & PG_r \\ * & -\Gamma_1 & 0 & 0 & 0 \\ * & * & -\Gamma_2 & 0 & 0 \\ * & * & * & . & 0 \\ * & * & * & * & 0 \\ * & * & * & * & -\Gamma_r \end{bmatrix} \quad (10.24)$$

where

$$M_0(s) = \tilde{A}_s^T P + P\tilde{A}_s + \sum_{i=1}^{r} \Gamma_i$$

The unidirectional derivative is negative if $M_s \prec 0$ holds $\forall s \in [1, .., \eta]$. By using (10.10), (10.15), and (10.24), inequality $M_s \prec 0$ is equivalent to (10.17). □

As given conditions of Theorem 10.2 are not useful for synthesis, the corollary below gives synthesis conditions for the stabilizing controllers. In fact, Theorem 10.2 is worked out for this purpose. The result of Theorem 10.2 can be used for the synthesis of the required controllers, as given by the following result.

Corollary 10.1 *If there exist symmetric matrices $X_1 \succ 0$, $X_2 \succ 0$, $\bar{Q}_1 \succ 0, \ldots, \bar{Q}_\eta \succ 0$, $\bar{R}_1 \succ 0, \ldots, \bar{R}_\eta \succ 0$, and matrices Y_1, Y_2, Z_1 and Z_2 and scalar $\rho > 0$ such that:*

$$\begin{bmatrix} [(\Xi)_{11} & (\Xi)_{12} & (G_1)_{11}X_1 & (G_1)_{12}X_2 & .. & (G_r)_{11}X_1 & (G_r)_{12}X_2 \\ * & (\Xi)_{22} & (G_1)_{21}X_1 & (G_1)_{22}X_2 & .. & (G_r)_{21}X_1 & (G_r)_{22}X_2 \\ * & * & -\bar{Q}_1 & 0 & 0 & 0 & 0 \\ * & * & * & -\bar{R}_1 & 0 & 0 & 0 \\ .. & .. & .. & .. & .. & .. & .. \\ * & * & * & * & * & -\bar{Q}_r & 0 \\ * & * & * & * & * & * & -\bar{R}_r \end{bmatrix} \prec 0 \quad (10.25)$$

$$\forall s \in [1, .., \eta]$$

$$\begin{bmatrix} \mu & (Z_1)_l & (Z_2)_l \\ * & X_1 & 0 \\ * & 0 & X_2 \end{bmatrix} \succ 0, \ \forall l \in [1, .., m] \quad (10.26)$$

hold true, then, the 2D saturated continuous system with multi-delays (10.14) is asymptotically stable.
where

$$(\Xi)_{11} = [A_{11}X_1 + B_1(D_sY_1 + D_s^- Z_1)]^{sym} + \sum_{i=1}^{\eta} \bar{Q}_i$$

$$(\Xi)_{12} = A_{12}X_2 + X_1A_{21}^T + B_1(D_sY_2 + D_s^- Z_2) + (D_sY_1 + D_s^- Z_1)^T B_2^T$$

$$(\Xi)_{22} = [A_{22}X_2 + B_2(D_sY_2 + D_s^- Z_2)]^{sym} + \sum_{i=1}^{\eta} \bar{R}_i$$

and $\mu = 1/\rho$, $Y_1 = K_1X_1$, $Y_2 = K_2X_2$, $Z_1 = H_1X_1$, $Z_2 = H_2X_2$, $X_1 = P_1^{-1}$, $X_2 = P_2^{-1}$, $Q_i = P_1\bar{Q}_iP_1$ and $R_i = P_2\bar{R}_iP_2$, for $i = 1, \ldots, \eta$.

Proof To obtain LMI (10.25), pre- and post-multiply inequality (10.17) by $diag\{P^{-1}, \ldots, P^{-1}\}$, while noting that $X = P^{-1}$, $Y = KX$, $Z = HX$ and $\bar{\Gamma}_i = X\Gamma_i X$. Then, we obtain the following LMI:

$$\Phi_s = \begin{bmatrix} \Phi_0(s) & G_1X & . & . & G_rX \\ * & -\overline{\Gamma}_1 & 0 & 0 & 0 \\ * & * & -\overline{\Gamma}_2 & 0 & 0 \\ * & * & * & * & 0 \\ * & * & * & * & -\overline{\Gamma}_r \end{bmatrix} \prec 0 \tag{10.27}$$

$$\forall s \in [1, \ldots, \eta]$$

with:

$$\Phi_0(s) = \Pi(s) + \Pi(s)^T + \sum_{i=1}^{r} \overline{\Gamma}_i \tag{10.28}$$

Let's note:

$$\begin{aligned} \Pi(s) = \tilde{A}_sP^{-1} &= AP^{-1} + B[D_sKP^{-1} + D_s^-HP^{-1}] \\ &= AX + B[D_sY + D_s^-Z] \end{aligned} \tag{10.29}$$

with:

$$Y = [Y_1\ Y_2],\ Z = [Z_1\ Z_2],\ K = [Y_1X_1^{-1}\ \ Y_2X_2^{-1}],$$

and

$$H = [Z_1X_1^{-1}\ \ Z_2X_2^{-1}] \tag{10.30}$$

By using conditions (10.15), (10.10), and (10.29), inequality (10.25) holds.
On the other hand, the inclusion condition (10.16) is equivalent to $\rho H_l P^{-1} H_l^T < 1$, which is equivalent to $\rho(HX)_l X^{-1}(XH)_l^T < 1$. Using (10.30), one then obtains:

$$\mu - Z_l X^{-1} Z_l^T > 0$$

By using the Schur complement, with $\mu = 1/\rho$, we have:

$$\begin{bmatrix} \mu & Z_l \\ * & X \end{bmatrix} \succ 0,\ l = 1, \ldots, m$$

Remember that $Z = [Z_1\ Z_2]$ and:

$$X = \begin{bmatrix} X_1 & 0 \\ 0 & X_2 \end{bmatrix},$$

then:

$$\begin{bmatrix} \mu & (Z_1)_l & (Z_2)_l \\ * & X_1 & 0 \\ * & 0 & X_2 \end{bmatrix} \succ 0, \; l = 1, ..., m$$

This ends the proof. □

Example 10.1 In order to show the applicability of our results, consider the 2D continuous system given by (10.1), where:

$$A_{11} = \begin{bmatrix} 1 & -0.5 \\ 0 & -2 \end{bmatrix}, A_{12} = \begin{bmatrix} 0.1 & -1 \\ 0 & 0.1 \end{bmatrix}$$
$$A_{21} = \begin{bmatrix} -1 & 0 \\ 0 & 0.1 \end{bmatrix}, A_{22} = \begin{bmatrix} 0 & -3 \\ 1 & -0.6 \end{bmatrix}$$
$$B_1 = \begin{bmatrix} 1 & -1 \\ 1 & 0 \end{bmatrix}, B_2 = \begin{bmatrix} 0 & 0 \\ 1 & 1 \end{bmatrix}.$$

The delay matrices are given by:

$$(G_1)_{11} = \begin{bmatrix} 0.30 & -0.15 \\ 0 & -0.60 \end{bmatrix}, (G_1)_{12} = \begin{bmatrix} 0.03 & -0.30 \\ 0 & 0.03 \end{bmatrix}$$
$$(G_1)_{21} = \begin{bmatrix} -0.30 & 0 \\ 0 & 0.03 \end{bmatrix}, (G_1)_{22} = \begin{bmatrix} 0 & -0.90 \\ 0.30 & -0.18 \end{bmatrix}$$
$$(G_2)_{11} = \begin{bmatrix} 0.10 & -0.15 \\ 0 & -0.20 \end{bmatrix}, (G_2)_{12} = \begin{bmatrix} 0.01 & -0.10 \\ 0 & 0.01 \end{bmatrix}$$
$$(G_2)_{21} = \begin{bmatrix} -0.1 & 0 \\ 0 & 0.01 \end{bmatrix}, (G_2)_{22} = \begin{bmatrix} 0 & -0.30 \\ 0.10 & -0.06 \end{bmatrix}$$

LMI's (10.25) and (10.26) are feasible with $\rho = 20$. The solution are given by:

$$K_1 = \begin{bmatrix} -4.3340 & 0.8914 \\ 5.7537 & -1.0677 \end{bmatrix}, \quad K_2 = \begin{bmatrix} 1.6379 & -4.8830 \\ 1.2071 & -5.1331 \end{bmatrix}$$
$$Q_1 = \begin{bmatrix} 0.0388 & 0.1022 \\ 0.1022 & 0.5801 \end{bmatrix}, \quad Q_2 = \begin{bmatrix} 0.0317 & 0.0885 \\ 0.0885 & 0.4902 \end{bmatrix}$$
$$R_1 = \begin{bmatrix} 0.0940 & 0.0302 \\ 0.0302 & 0.0463 \end{bmatrix}, \quad R_2 = \begin{bmatrix} 0.0538 & 0.0098 \\ 0.0098 & 0.0274 \end{bmatrix}$$
$$P_1 = \begin{bmatrix} 30.3924 & -4.9734 \\ -4.9734 & 3.0245 \end{bmatrix}, \quad P_2 = \begin{bmatrix} 9.2521 & -7.9980 \\ -7.9980 & 28.4021 \end{bmatrix}$$
$$H_1 = \begin{bmatrix} -0.9616 & 0.1525 \\ 1.1472 & -0.2154 \end{bmatrix}, \quad H_2 = \begin{bmatrix} 0.3638 & -0.8667 \\ 0.0659 & -1.0368 \end{bmatrix}$$

For simulations purpose, we have used the sampling periods $T_h = 0.1$ and $T_v = 0.08$. The delays are $\tau_1^h = 1$, $\tau_2^h = 2$, $\tau_1^v = 1$, and $\tau_2^v = 2$.

Hence, the stabilizing state feedback is easily obtained. Figs. 10.1 and 10.2 present the evolution of the components of the states $x^h(t_1, t_2)$ and $x^v(t_1, t_2)$. Figure 10.3 presents the evolution of control components $u_1(t_1, t_2)$ and $u_2(t_1, t_2)$. The stability domains are represented in Fig. 10.4.

It is clear that the 2D saturated system with multiple delays is asymptotically stable and converges toward the origin while allowing saturation on the control.

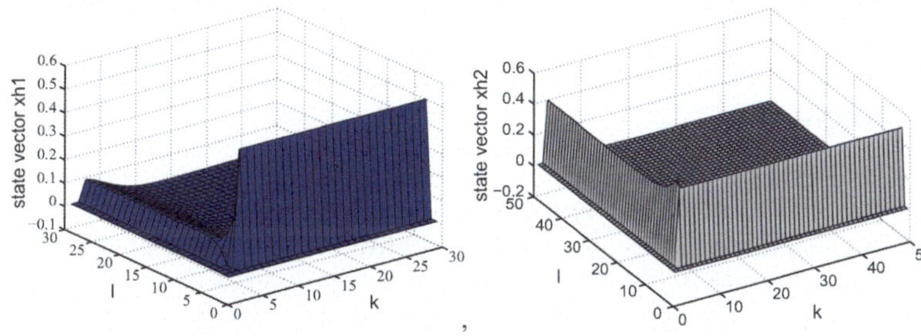

Fig. 10.1 States evolution of the components of $x^h(t_1, t_2)$

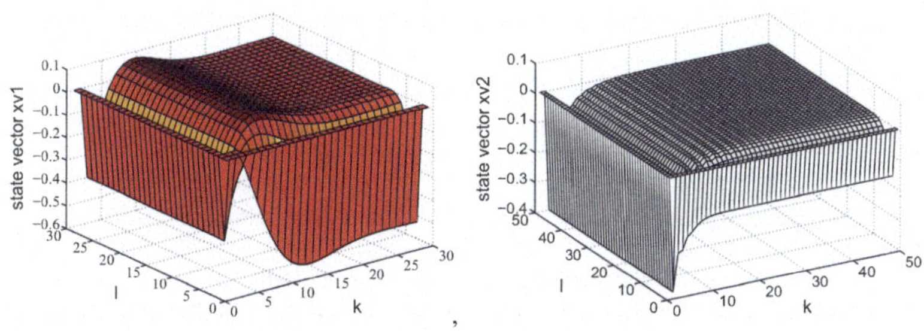

Fig. 10.2 States evolution of the components of $x^v(t_1, t_2)$

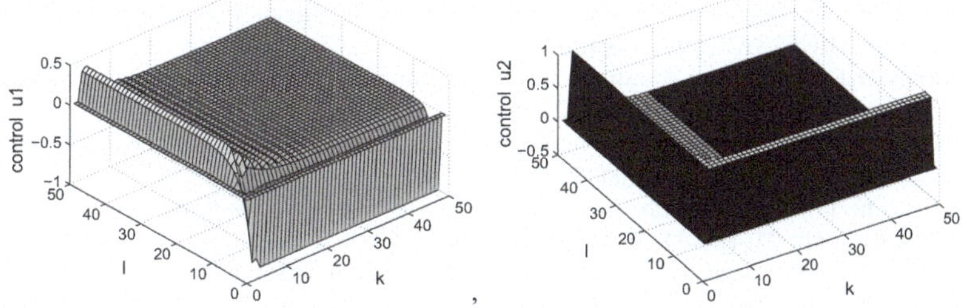

Fig. 10.3 Evolution of the components of control $u_1(t_1, t_2)$ and $u_2(t_1, t_2)$

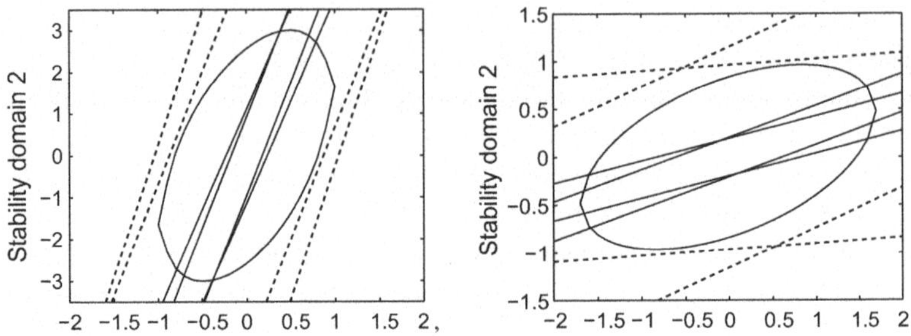

Fig. 10.4 Stability domains

10.5 Conclusion

In this chapter, the problem of stabilizability of the 2D saturated continuous systems with multi-delays is studied. State feedback control is used. Sufficient conditions of asymptotic stability are derived. The synthesis of the required controllers is given under LMI form. Numerical example is provided to illustrate the result.

References

1. M. Benhayoun, Contribution à la commande des systèmes 2D retardés avec contraintes sur la commande. Cadi Ayyad University, thèse de doctorat d'état, 2011
2. M. Benhayoun, F. Mesquine, A. Benzaouia, Delay-dependent Stabilizability of 2D delayed continuous systems with saturating control. Circuits Syst. Signal Process. (CSSP) **32**(2), 2723–2743 (2013)
3. A. Benzaouia, A. Hmamed, F. Mesquine, M. Benhayoun, F. Tadeo, Stabilization of 2D discrete time saturated systems by state feedback control, in *Conference STA* (Tunisia, 2009), Dec. 20–22
4. M. Benhayoun, A. Benzaouia, F. Mesquine, F. Tadeo, Stabilization of 2D continuous systems with multi-delays and saturated control, in *18th Mediterranean Conference on Control and Automation* (Marrakech, Morocco, 2010), June 23–25
5. A. Benzaouia, M. Benhayoun, F. Tadeo, State feedback stabilization of 2D continuous systems with delays. IJICIC **7**(2), 977–988 (2011)
6. E. Fornasini, G. Marchesini, Doubly-indexed dynamical systems: state-space models and structural properties. Math. Syst. Theory **12**, 59–72 (1978)
7. D.D. Givone, R.P. Roesser, Multidimentional linear iterative circuits-general properties. IEEE Trans. Comp. **21**(10), 1067–1073 (1972)
8. E. Fornasini, G. Marchesini, State-space realization theory of two-dimentional filters. IEEE Trans. Aut. Contr. **21**(4), 484–492 (1976)
9. K. Galkowski, LMI based stability analysis for 2D contiuous systems. Int. Conf. Electr. Circuits Syst. **3**, 923–926 (2002)
10. A. Hmamed, A. Benzaouia, H. Bensalah, Regulator problem for linear continuous time-delay systems with non symmetrical constrained control. IEEE Tran. Autom. Control **40**, 1615–1619 (1995)
11. A. Hmamed, Constrained regulation of linear discrete-time systems with time-delay: delay-dependent and delay independent conditions. Int. J. Syst. Sci. **31**, 529–536 (2000)

12. A. Hmamed, F. Mesquine, F. Tadeo, M. Benhayoun, A. Benzaouia, Stabilization of 2D saturated systems by state feedback control. Multidim. Syst. Signal Process. **21**(3), 277–292 (2010)
13. T. Kaczorek, *Two Dimensional Linear Systems* (Springer, 1985)
14. T. Kaczorek, Realization problem, reachability and minimum energy control of positive 2D Roesser model, in *Proceedings of the 6th Annual International conferences Advances in communication and control* (1997), pp. 765–776
15. W. Marszalek, Two dimensional state-space discrete models for hyperbolic partial differential equations. Appl. Math. Models **8**, 11–14 (1984)
16. S.I. Niculescu, *Delay Effects on Stability. A Robust Control Approach*. Lecture Notes in Control and Information Science (Springer, Heidelberg, 2001)
17. J.W. Lam, K. Galkowski, S. Xu, Z. Lin, Robust stability and stabilization of 2d discrete state-delayed systems. Syst. Control Lett. **51**, 277–291 (2004)
18. W. Paszke, J. Lam, K. Galkowski, S. Xu, Z. Lin, E. Rogers, A. Kummert, Delay-dependant stability of 2-D state-delayed linear systems, in *IEEE ISCAS* (2006), pp. 2813–2816
19. M. Piekarski, Algebraic characterization of matrices whose multivariable characteristic polynomial is hermitian, in *Proceedings of the International Symposium Operator Theory* (Lubbock, TX, 1977), pp. 121–126
20. R. Roesser, A discrete state-space model for linear image processing. IEEE Trans. Autom. Contr. AC **20**, 1–10 (1975)

Chapter 11
Case Studies

11.1 Introduction

Case studies or application of theoretical results to real life processes is of great importance for researchers. Effectively, the aim of all theoretic works and new results is to show applicability and effectiveness in real world processes. Further, the validation of the obtained results for general modeling and general theory is upon application. The developed work in this book that was devoted in general to constrained control and/or state systems and different problems that arise in such systems cases, may be also validated through application on some real processes. Henceforth, this chapter is devoted to the present application of our previous results to a number of real systems. The focus will be on rising of constraints in such systems, and the treatment that will be used accordingly. First, the control of the pH degree in a stirred tank is obtained. The process is modeled as an uncertain system with constrained control. The application of the previously presented results about robust-constrained control is successively obtained. Second, the wastewater treatment plant more precisely the activated sludge process is considered. It will be controlled using the positive invariance techniques. This process emphasizes the use of observers as the states, say concentrations, is not measurable, and the sensors for such processes are very expensive and may not work on line when directing the process.

11.2 Application to a pH-process

Consider the pH-control plant shown in Fig. 11.1 (already presented in [27]), [18] which consists of a stirred tank where a solution of high concentration of the acid ClH is mixed with water to obtain a liquid of controlled pH. The mixture's pH is measured using a pH-meter (Kent EIL9143), which presents appreciable inertia. Water is fed from a tank using a peristaltic pump, which produces a variable flow depending on the level of the liquid in the tank (Fig. 11.1). Variations in the dynamics due to

A. Benzaouia et al., *Saturated Control of Linear Systems*, Studies in Systems, Decision and Control 124, https://doi.org/10.1007/978-3-319-65990-9_11

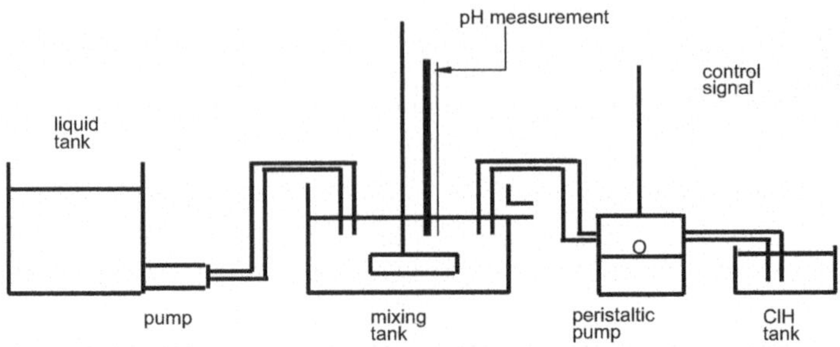

Fig. 11.1 Laboratory Plant

changing flows, concentrations, and operating points make this system uncertain. It can be seen in Fig. 11.1 that the control scheme in this plant is based on acid flow control. However, the techniques presented now could be directly applied to other control structures such as base flow control [26], base flow control in the presence of buffer flow [6] or base and acid flow control [9].

Although the modeling of pH-control processes is well studied [10], in our case, it is only necessary to have a simplified model, based on first principles. Assuming that the input liquid is pure water that the acid solution has constant concentration on ClH, and there is perfect solution, mixing, and no buffering; the following model can be obtained [23]:

$$M\frac{dN_d}{dt} = -q_o N_d - q_a N_d + q_a N_a$$

$$\varsigma\frac{dN_d^*}{dt} = N_d - N_d^*$$

$$pH = -log_{10}(N_d^*)$$

where ς is the measurement time constant, M is the mass of the liquid in the tank, q_a is the acid mass flow, q_o is the liquid mass flow, N_d is the acid concentration in the tank, N_d^* is the measured concentration, and N_a is the input acid concentration. The objective of the control system is to maintain the pH of the liquid in the mixing tank on desired values, using the acid mass flow (u) as the control variable. The model parameters were estimated using measured data.
The system can be represented by the following transfer function

$$G(s) = \frac{k}{(s-a)(s-b)},$$

where s presents here the Laplace variable and the gain k, the pole a is uncertain and the pole b is supposed to be constant. Uncertainty in the plant was experimentally estimated: different experiments under the most extreme conditions were done (maximum and minimum flows, extreme values of pH, etc.) then the parameters were

measured, using additional sensors where it was necessary (e.g., to evaluate the inlet flow range). The model was then linearized at these extreme points and the extreme variations of gain and time constant selected as uncertainty in the model.
The system can be represented in state space by:

$$A = \begin{bmatrix} 0 & 1 \\ -ba & b+a \end{bmatrix} \text{ and } B = \begin{bmatrix} 0 \\ k \end{bmatrix}$$

with constrained control u in Ω given by:

$$\Omega = \left\{ u \in \mathbb{R}^m / -u_{min} \le u \le u_{max},\ u_{max},\ u_{min} \in \mathbb{R}^m_+ \right\}.$$

where $u_{\max} = 60$, and $u_{\min} = 35$.

It is experimentally known that the parameter b can be assumed to be constant ($b = -0.012725$), as it suffers small variations which do not affect significantly the plant behavior. On the other hand, parameters a and k change greatly between working points, so they represent the parametric uncertainty. The parameter variations experimentally obtained were:

$$k \in [-0.4649\ 10^{-4}, -0.7449\ 10^{-4}]$$
$$a \in [-0.25, -2]$$

Let

$$a_o = \frac{a_{\max} + a_{\min}}{2}, \qquad k_o = \frac{k_{\max} + k_{\min}}{2},$$

then any $a \in [a_{\min}, a_{\max}]$, $k \in [k_{\min}, k_{\max}]$ can be written as:

$$a = a_o + q_A, \qquad k = k_o + q_B$$

where:

$$-\Delta a \le q_A \le \Delta a, \qquad -\Delta k \le q_B \le \Delta k$$

with:

$$\Delta a = \frac{a_{\max} - a_{\min}}{2}, \qquad \Delta k = \frac{k_{\max} - k_{\min}}{2}.$$

So, the form of the system becomes as (1.27), with:

$$A(q_A(t)) = \begin{bmatrix} 0 & 1 \\ -b(a_o + q_A) & b + a_o + q_A \end{bmatrix} \text{ and } B(q_B(t)) = \begin{bmatrix} 0 \\ k_o + q_B \end{bmatrix}$$

The vertices of uncertainties domain are given by

$$\nu_{11} = (0.875, 1.41\ 10^{-5}); \qquad \nu_{12} = (0.875, -1.41\ 10^{-5})$$
$$\nu_{21} = (-0.875, 1.41\ 10^{-5}); \qquad \nu_{22} = (-0.875, -1.41\ 10^{-5}).$$

The nominal system can be given by (1.29) where

$$A_o = \begin{bmatrix} 0 & 1 \\ -ba_o & b + a_o \end{bmatrix} \text{ and } B_o = \begin{bmatrix} 0 \\ k_o \end{bmatrix}$$

Therefore, to assign two new eigenvalues to the closed-loop system, and without loss of generality, the system is augmented with fictitious constrained input v. This augmented system is given by

$$\dot{x}(t) = A_o x(t) + B_{oa} \begin{bmatrix} u \\ v \end{bmatrix}$$

B_{oa} is the matrix B_o augmented by $(n - m)$ null columns i.e.:

$$B_{oa} = \begin{bmatrix} 0 & 0 \\ k_0 & 0 \end{bmatrix}$$

Constraints on the fictitious control v are given by $v_{\max} = 40$ and $v_{\min} = 50$.

In order to design a stabilizing controller for the nominal system, we use the so-called inverse procedure [4] given by Theorem 1.6. For this, the matrix

$$H_o = \begin{bmatrix} -0.0130 & -0.0070 \\ 0 & -1.1245 \end{bmatrix}$$

is selected satisfying conditions (1.24) and (1.31), then the algebraic equation $XA_o + XB_{oa}X = H_oX$ is solved. Solution of this equation leads to the following augmented regulator

$$F_a = 10^3 \begin{bmatrix} 0.0050 & -0.0037 \\ -0.0170 & -1.3107 \end{bmatrix}$$

Matrices $H(v_{ij})$ for $i = 1, 2. j = 1, 2$ can be calculated as solutions of the equations $F_a A(v_i) + F_a B_a(v_j) F_a = X(v_{ij}) F_a$. It is easy to check that condition (2.22), i.e.,

$$\tilde{H}_c(v_{ij}) U_a \leq 0 \text{ where } U_a = [u_{max}^T \; v_{max}^T \; u_{min}^T \; v_{min}^T]^T$$

is satisfied for all $v_{ij}, i = 1, 2; j = 1, 2$. Hence, the computed regulator gain F_a is robust. Notice here that for the effective control, the feedback signal can be extracted from F_a as

$$F = [5 \quad -3.7]$$

Figures 11.2, 11.3, 11.4, and 11.5 show the state vector evolution form initial condition $x_o = [0.02 \; 0.03]^T$, the control evolution and the evolution of the uncertain parameters a and k in time. It can be seen that robust regulation is achieved with

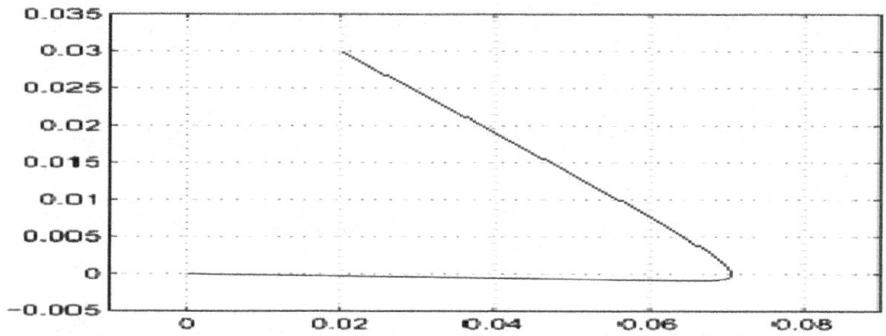

Fig. 11.2 Evolution of the state vector

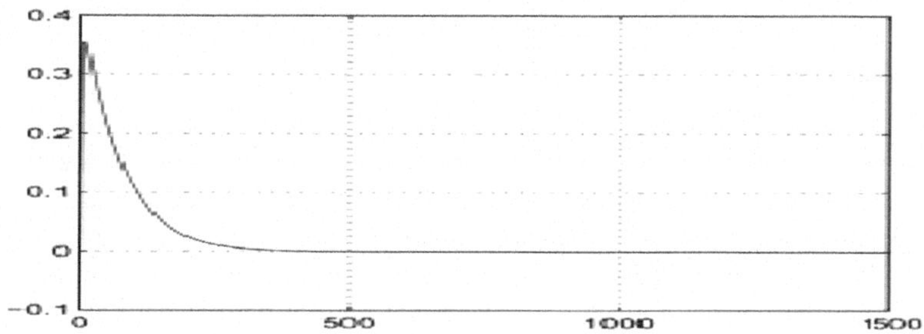

Fig. 11.3 Control evolution

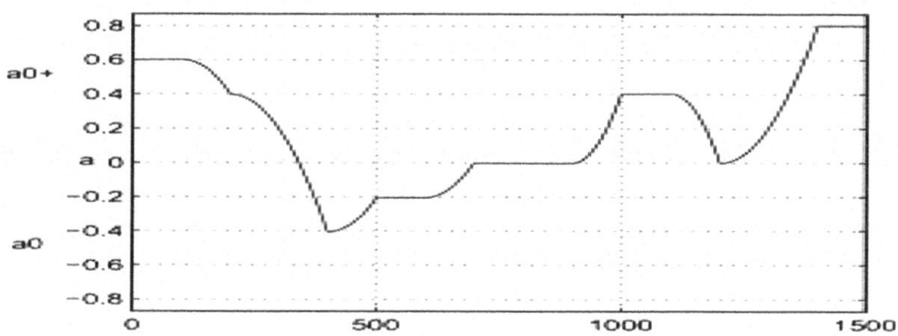

Fig. 11.4 Evolution of the pole a

respected control constraints. It can be seen that the theoretical results established in Chap. 2 are successfully applied to the real process consisting of the control of the pH degree in a mixed stirred tank. This application is of great interest in some industries, like sugar production, where the pH is very important for the products. Modeling uncertainties and constrained control were handled in this application showing the effectiveness of the proposed approach.

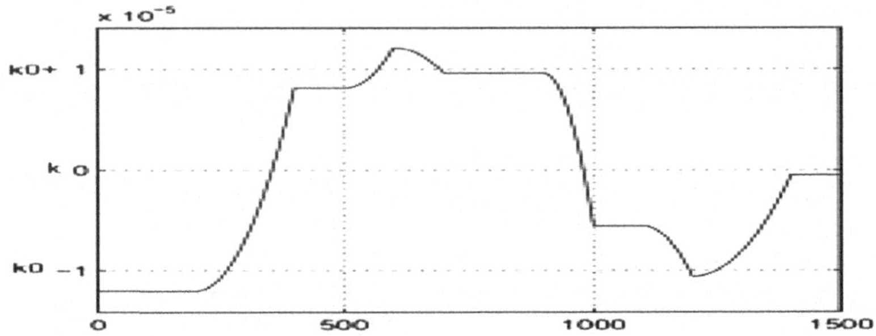

Fig. 11.5 Variation of coefficient k

11.3 Wastewater Treatment Plant (WWTP)

Our second application is a wastewater treatment plant. Wastewater treatment is just one component in the urban water cycle. However, it is an important one since it ensures that the environmental impact of water human usage is significantly reduced. The modeling and control of activated sludge process, which are recognized as the most common and major unit process for reduction of organic waste, have become a subject of great interest. Researchers [2, 8, 19, 20, 28] have investigated different control strategies for the monitoring of such processes. The development of effective control strategies on this kind of Wastewater treatment plants is hampered by the inherent nonlinearities, the time-varying dynamics and the lack of suitable instrumentation.

The state space representation is frequently used to form multivariable approach to linear control. The most common control schemes are based on availability of the state for feedback. In same real process, it is either impossible or inappropriate to measure all elements of the system state. To overcome this problem, an auxiliary dynamical system, known as observer, driven by the inputs and outputs of the original system is designed [13].

As pointed out along the chapters of this book, another problem that arises when considering real process is the limitation of state or control of the process. In fact, processes are naturally nonlinear and to obtain linear useful model, approximation of small variations around the steady state are used. Hence, validity of such linear model is limited to a neighborhood of the steady point leading to constraints on some variables. Further, inherent physical limitations may be the source of limited variables. The respect of these constraints can be accomplished by designing suitable feedback control laws. In many cases, this can be done by constructing positively invariant domains inside the set of the constraints ([1, 4, 5, 17]). Other important applications were derived from this concept. In particular, the observers in the framework of positive invariant sets is given in [15, 16].

During the last decades, many investigations have been focused on the control of the nitrogen and dissolved oxygen in an activated sludge reactor within a WWTP with different strategies. One may quote predictive control, optimal control, and adaptive control ... [3, 22]. Note here that constraints on the control are not handled, and further, all required measurements are assumed available. Apart from this, one may also cite works about the same topic but limited to estimation [7] and not the control. Furthermore, works combining estimation to control for monitoring such processes can also be found [25]. However, constraints are not taken into account during the design steps. Therefore, this works may be thought as a generalization where constraints, estimation, and control are considered using the positive invariance concept together at the design stage.

The objective here is to apply positive invariance concept techniques to a WWTP. The obtained linearized model combines the problems of non-availability of the state to measure with the limitations of some variables. The control is achieved by an observer-based controller that can take into account constraints on the control and on the error. The obtained linear model is worked out to meet all design required conditions. The efficiency of the process monitoring is showed via simulations with the real plant.

11.3.1 Process Modeling

A typical, conventional activated sludge plant for the removal of carbonaceous and nitrogen materials consists of an anoxic basin followed by an aerated one and a settler (see the Fig. 11.6). In the presence of dissolved oxygen, wastewater, that is mixed with the returned activated sludge, is biodegraded in the aerated reactor. Treated effluent is separated from the sludge and wasted while a large fraction is returned to the anoxic reactor to maintain an appropriate substrate to biomass ratio. In this work, six basic components are present in the wastewater: autotrophic bacteria X_A, heterotrophic bacteria X_H, readily biodegradable carbonaceous substrates S_S, nitrogen substrates S_{NH}, S_{NO} and dissolved oxygen S_O where X_A, X_H, S_S, S_{NH}, S_{NO}, and S_O represent the concentrations of these elements. In the modeling of the process, the following assumptions are considered: first, the physical properties of the fluid are constant and there is no concentration gradient across the vessel.

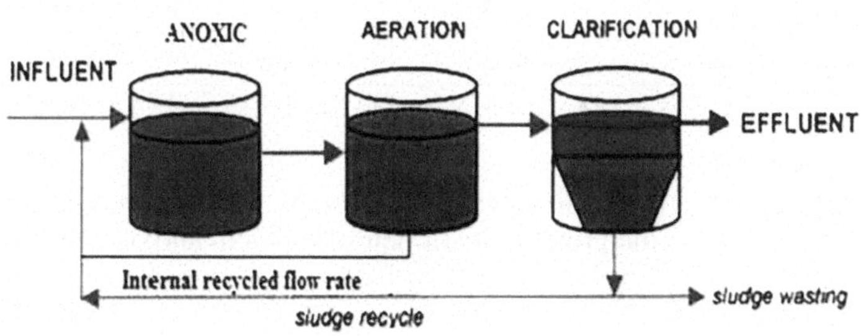

Fig. 11.6 W.W.T. Plant

Second, substrates and dissolved oxygen are considered as rate limiting with a bi-substrate Monod-type Kinetic. Finally, no bioreaction takes place in the settler that is considered perfect.

Based on the above description and assumptions, the full set of ordinary differential equations (mass balance equations) making up the IAWQ (ASM1) Model NO.1 is obtained [11, 21].

11.3.1.1 Modeling of the Aerated Basin

In the aerated basin, writing the mass balance equations lead to the following:

$$\dot{X}_{A,nit}(t) = (1 + r_1 + r_2)\, D_{nit} \left(X_{A,denit} - X_{A,nit}\right) + \left(\mu_{A,nit} - b_A\right) X_{A,nit} \tag{11.1}$$

$$\dot{X}_{H,nit}(t) = (1 + r_1 + r_2)\, D_{nit} \left(X_{H,denit} - X_{H,nit}\right) + \left(\mu_{H,nit} - b_H\right) X_{H,nit} \tag{11.2}$$

$$\dot{S}_{S,nit}(t) = (1 + r_1 + r_2)\, D_{nit} \left(S_{S,denit} - S_{S,nit}\right) + \left(\mu_{H,nit} + \mu_{Ha,nit}\right) X_{H,nit}/Y_H \tag{11.3}$$

$$\begin{aligned}\dot{S}_{NH,nit}(t) = {} & (1 + r_1 + r_2)\, D_{nit} \left(S_{NH,denit} - S_{NH,nit}\right) + (i_{xb} + 1/Y_A)\, \mu_{A,nit} X_{A,nit} \\ & - \left(\mu_{H,nit} + \mu_{Ha,nit}\right) i_{xb} X_{H,nit}\end{aligned} \tag{11.4}$$

$$\begin{aligned}\dot{S}_{NO,nit}(t) = {} & (1 + r_1 + r_2)\, D_{nit} \left(S_{NO,denit} - S_{NO,nit}\right) + \mu_{A,nit} \tfrac{X_{A,nit}}{Y_A} \\ & - \tfrac{1-Y_H}{2.86 Y_A} \mu_{Ha,nit} X_{H,nit}\end{aligned} \tag{11.5}$$

$$\begin{aligned}\dot{S}_{O,nit}(t) = {} & (1 + r_1 + r_2)\, D_{nit} \left(S_{O,denit} - S_{O,nit}\right) + a_o Q_{air}(C_S - S_{O,nit}) \\ & - \tfrac{4.57 - Y_A}{Y_A} \mu_{A,nit} X_{A,nit} - \tfrac{1-Y_H}{Y_H} \mu_{Ha,nit} X_{H,nit}\end{aligned} \tag{11.6}$$

where:

$$\mu_{A,nit} = \mu_{max,A} \frac{S_{NH,nit}}{\left(K_{NH,A} + S_{NH,nit}\right)} \frac{S_{O,nit}}{(K_{O,A} + S_{O,nit})}$$

$$\mu_{H,nit} = \mu_{max,H} \frac{S_{S,nit}}{(K_S + S_{S,nit})} \frac{S_{NH,nit}}{\left(K_{NH,H} + S_{NH,nit}\right)} \frac{S_{O,nit}}{(K_{O,H} + S_{O,nit})}$$

$$\mu_{Ha,nit} = \mu_{max,H} \frac{S_{S,nit}}{(K_S + S_{S,nit})} \frac{S_{NH,nit}}{\left(K_{NH,H} + S_{NH,nit}\right)} \frac{K_{O,H}}{(K_{O,H} + S_{O,nit})} \frac{S_{NO,nit}}{(K_{NO} + S_{NO,nit})} \theta_{NO}$$

$\mu_{A,nit}$ and $\mu_{H,nit}$ are the growth rates of autotrophs and heterotrophs in aerobic conditions and $\mu_{Ha,nit}$ is the growth rate of heterotrophs in anoxic conditions.

11.3.1.2 Modeling of the Anoxic Basin

In the anoxic basin, mass balance equations lead to the following:

$$\dot{X}_{A,denit}(t) = D_{denit}(X_{A,in} + r_1 X_{A,nit}) - (1 + r_1 + r_2) D_{denit} X_{A,denit} + \\ + \alpha . r_2 D_{denit} X_{rec} + (\mu_{A,denit} - b_A) X_{A,denit} \tag{11.7}$$

$$\dot{X}_{H,denit}(t) = D_{denit}(X_{H,in} + r_1 X_{H,nit}) - (1 + r_1 + r_2)\, D_{denit} X_{H,denit} + \\ + (1 - \alpha) r_2 D_{denit} X_{rec} + (\mu_{H,denit} - b_H) X_{H,denit} \tag{11.8}$$

$$\dot{S}_{S,denit}(t) = -(\mu_{H,denit} + \mu_{Ha,denit}) \frac{X_{H,denit}}{Y_H} - (1 + r_1 + r_2)\, D_{denit} S_{S,denit} \\ + D_{denit}(S_{S,in} - r_1 S_{S,nit}) \tag{11.9}$$

$$\dot{S}_{NH,denit}(t) = D_{denit}(S_{NH,in} - r_1 S_{NH,nit}) - (1 + r_1 + r_2)\, . D_{denit} S_{NH,denit} \\ - (i_{xb} + 1/Y_A)\, \mu_{A,denit} X_{A,denit} - \left(\mu_{H,denit} + \mu_{Ha,denit}\right) i_{xb} X_{H,denit} \tag{11.10}$$

$$\dot{S}_{NO,denit}(t) = D_{denit}(S_{NO,in} - r_1 S_{NO,nit}) - (1 + r_1 + r_2)\, D_{denit} S_{NO,denit} + \\ + \frac{\mu_{A,denit} X_{A,denit}}{Y_A} - \frac{1 - Y_H}{2.86 Y_H} \mu_{Ha,denit} X_{H,denit} \tag{11.11}$$

where:

$$\mu_{A,denit} = \mu_{max,A} \cdot \frac{S_{NH,denit}}{\left(K_{NH,A} + S_{NH,denit}\right)}$$

$$\mu_{H,denit} = \mu_{max,H} \cdot \frac{S_{S,denit}}{(K_S + S_{S,denit})} \cdot \frac{S_{NH,denit}}{\left(K_{NH,H} + S_{NH,denit}\right)}$$

$$\mu_{Ha,denit} = \mu_{max,H} \cdot \frac{S_{S,denit}}{(K_S + S_{S,denit})} \cdot \frac{S_{NH,denit}}{\left(K_{NH,H} + S_{NH,denit}\right)} \cdot \\ \frac{S_{NO,denit}}{(K_{NO} + S_{NO,denit})} . \theta_{NO}$$

11.3.1.3 Modeling of the Settler

In the settler, the mass balance equations enable us to write:

$$\dot{X}_{rec} = (1 + r_2) D_{dec}(X_{A,nit} + X_{H,nit}) - (r_2 + w) D_{dec} X_{rec} \tag{11.12}$$

Above, r_1, r_2, and w represent, respectively, the ratios of the internal recycled flow Q_{r1}, the recycled flow Q_{r2}, and the waste flow Q_w to influent flow Q_{in}. That is $Q_{r1} = r_1 Q_{in}$, $Q_{r2} = r_2 Q_{in}$, and $Q_w = w Q_{in}$. Further, C_S is the maximum dissolved oxygen concentration, and X_{rec} is the concentration of the recycled biomass. Finally, $D_{nit} = \frac{Q_{in}}{V_{nit}}$, $D_{denit} = \frac{Q_{in}}{V_{denit}}$, and $D_{dec} = \frac{Q_{in}}{V_{dec}}$ where D_{nit}, D_{denit}, and D_{dec} are the dilution rates in, respectively, nitrification, denitrification, basins, and settler tank. All remaining involved variables and parameters of the system (11.1)–(11.12) have been directly taken from the literature [24] and are defined in Tables 11.1 and 11.2. To obtain a model in the state space, the state vector is considered as follows:

Table 11.1 Process characteristics

Variable	Value	Description
V_{nit}	1333 m^3	Volume of nitrification basin
V_{denit}	1000 m^3	Volume of denitrification basin
V_{dec}	6000 m^3	Volume of of settler
Q_{in}	18446 m^3/j	Influent flow rate
Q_w	385 m^3/j	Waste flow rate
$X_{A,in}$	0 mg/l	Autotrophs in the influent
$X_{H,in}$	30 mg/l	Hetertrophs in the influent
$S_{S,in}$	200 mg/l	Substrate in the influent
$S_{NH,in}$	30 mg/l	Ammonium in the influent
$S_{NO,in}$	2 mg/l	Nitrate in the influent
$S_{O,in}$	0 mg/l	Oxygen in the influent

Table 11.2 Kinetic parameters and stoichiometric coefficient characteristic

Variable	Value	Description
Y_A	0.24	Yield of autotroph mass
Y_H	0.67	Yield of heterotroph mass
i_{xb}	0.086	
K_S	20 mg/l	Affinity constant
$K_{NH,A}$	1 mg/l	Affinity constant
$K_{NH,H}$	0.05 mg/l	Affinity constant
K_{NO}	0.5 mg/l	Affinity constant
$K_{O,A}$	0.4 mg/l	Affinity constant
$K_{O,H}$	0.2 mg/l	Affinity constant
μ_{Amax}	0.8 l/j	Maximum specific growth rate
μ_{Hmax}	0.6 l/j	Maximum specific growth rate
b_A	0.2 l/j	Decay coefficient of autotrophs
b_H	0.68 l/j	Decay coefficient of heterotrophs
θ_{NO}	0.8 l/j	Correction factor for anoxic growth
α	0.5	

$$X(t) = [X_{A,nit}(t)\ X_{H,nit}(t)\ S_{S,nit}(t)\ S_{NH,nit}(t)\ S_{NO,nit}(t)\ S_{O,nit}(t)\ X_{A,denit}(t)\ X_{H,denit}(t)\ S_{S,denit}(t)\ S_{NH,denit}(t)\ S_{NO,denit}(t)\ X_{rec}(t)]^T. \tag{11.13}$$

Further, to complete the model, the following input and output vectors are used

$$U = [Q_{r1}\ \ Q_{r2}\ \ Q_{air}]^T, \tag{11.14}$$

$$Y(t) = [S_{NH,nit}(t)\ \ S_{NO,nit}(t)\ \ S_{O,nit}(t)]^T \tag{11.15}$$

The constraints on the control are given by the following limitations:

$$\begin{cases} -\bar{Q}_{r1} \leq Q_{r1} \leq 4\bar{Q}_{r1} \\ -\bar{Q}_{r2} \leq Q_{r2} \leq \bar{Q}_{r2} \\ -\bar{Q}_{air} \leq Q_{air} \leq 2\bar{Q}_{air} \end{cases} \tag{11.16}$$

Linearizing the system around the equilibrium point computed from the nonlinear equations leads to the new variables (x, u, y) that are now deviation variables. That is they are deviations from the point the model is linearized about, not their original absolute values. The equilibrium point is given by:

$$\bar{x} = [69.6 \;\; 623 \;\; 13.5 \;\; 3.2 \;\; 10.4 \;\; 2.4 \;\; 68.9 \; 624.6 \;\; 20.9 \;\; 8.9 \;\; 5.3 \;\; 1356.8]^T \tag{11.17}$$

and $\bar{Q}_{r1} = 2300\ m^3/j$, $\bar{Q}_{r2} = 18446\ m^3/j$ and $\bar{Q}_{air} = 100\ m^3/j$ which leads to the following matrices for the system:

$$A = \begin{bmatrix}
-29.07 & 0 & 0 & 2.65 & 0 & 2.17 & 29.40 & 0 & 0 & 0 & 0 & 0 \\
0 & -29.48 & 6.04 & 0.64 & 0 & 4.40 & 0 & 29.40 & 0 & 0 & 0 & 0 \\
0 & -0.34 & -38.99 & -1.02 & -0.05 & 5.00 & 0 & 0 & 29.40 & 0 & 0 & 0 \\
-2.22 & -0.02 & -0.55 & -40.74 & 0 & 12.23 & 0 & 0 & 0 & 29.40 & 0 & 0 \\
2.18 & 0 & -0.06 & 11.05 & -29.40 & 9.64 & 0 & 0 & 0 & 0 & 29.40 & 0 \\
-9.45 & 0 & -0.18 & -47.90 & -0.01 & -167.00 & 0 & 0 & 00 & 0 & 0 & \\
2.30 & 0 & 0 & 0 & 0 & 0 & -38.67 & 0 & 0 & 0.55 & 0 & 1.84 \\
0 & 2.30 & 0 & 0 & 0 & 0 & 0 & -39.18 & 4.44 & 0.11 & 0 & 16.60 \\
0 & 0 & 2.30 & 0 & 0 & 0 & 0 & -0.78 & -50.67 & -0.30 & -3.42 & 0 \\
0 & 0 & 0 & 2.30 & 0 & 0 & -3.06 & -0.04 & -0.66 & -39.32 & -0.19 & 0 \\
0 & 0 & 0 & 0 & 2.30 & 0 & 2.99 & -0.03 & -0.55 & 1.13 & -39.58 & 0 \\
6.14 & 6.14 & 0 & 0 & 0 & 0 & 0 & 0 & 0 & 0 & 0 & -3.13
\end{bmatrix}$$

$$B = 10^4 \begin{bmatrix} -0.0011 & -0.0011 & 0 \\ 0.0023 & 0.0023 & 0 \\ 0.0102 & 0.0102 & 0 \\ 0.0079 & 0.0079 & 0 \\ -0.0071 & -0.0071 & 0 \\ -0.0033 & -0.0033 & 0.0008 \\ 0.0014 & 0.1233 & 0 \\ -0.0031 & 1.1003 & 0 \\ -0.0386 & -0.0386 & 0 \\ -0.0105 & -0.0165 & 0 \\ 0.0094 & -0.0097 & 0 \\ 0 & -0.2042 & 0 \end{bmatrix} ; \quad C = \begin{bmatrix} 0\,0\,0\,1\,0\,0\,0\,0\,0\,0\,0\,0 \\ 0\,0\,0\,0\,1\,0\,0\,0\,0\,0\,0\,0 \\ 0\,0\,0\,0\,0\,1\,0\,0\,0\,0\,0\,0 \end{bmatrix}$$

Remark 11.1 It is worth noting here that the obtained state space representation is not controllable nor observable. In fact, matrices of controllability and observability are, respectively, of rank 10 and 9. Further, the matrix A of the system has a spectrum that contains stable eigenvalues, let say $n - m = 9$ stable eigenvalues with $x \in \mathbb{R}^n$, $u \in \mathbb{R}^m$ $y \in \mathbb{R}^{p_o}$.

11.3.1.4 Working Out the Model

Any representation in the state space can be transformed into the equivalent form by using the transformation $z = M\ x$ [12]:

$$\begin{cases} \dot{z} = \bar{A}\ z + \bar{B}\ u \\ y(t) = \bar{C}\ z \end{cases} \tag{11.18}$$

with:

$$\bar{A} = M\ A\ M^{-1} \quad \bar{B} = M\ B \quad \text{and } \bar{C} = C\ M^{-1}$$

$$\bar{A} = \begin{bmatrix} A_{c\bar{o}} & A_{12} \\ 0 & A_{co} \end{bmatrix}, \quad \bar{B} = \begin{bmatrix} B_{c\bar{o}} \\ B_{co} \end{bmatrix}, \quad \bar{C} = \begin{bmatrix} 0 & C_{co} \end{bmatrix} \quad \text{and } z = \begin{bmatrix} z_{c\bar{o}} \\ z_{co} \end{bmatrix}$$

Hence, the system may be re-written as follows:

$$\begin{cases} \dot{z}_{c\bar{o}} = A_{c\bar{o}} z_{c\bar{o}} + A_{12} z_{co} + B_{c\bar{o}} u \\ \dot{z}_{co} = A_{co} z_{co} + B_{co} u \\ y = C_{co} z_{co} \end{cases} \tag{11.19}$$

where (A_{co}, B_{co}, C_{co}) is controllable and observable. Further, computing the spectrum $\sigma(A_{c\bar{o}}) = \{-0.3373, -35.9885 + 1.2899i, -35.9885 - 1.2899i\}$ shows that it is stable, and hence stabilizing the system matrix A_{co} suffices to stabilizes the hole system [12]. For the WWTP, the states $(S_{NH,nit}(t) \quad S_{NO,nit}(t) \quad S_{O,nit}(t))$ are measurable, so the matrix M is chosen like:

$$
M = \left[\begin{array}{cccccccccccc}
0.0001 & 0.9975 & -0.0395 & 0 & 0 & 0 & 0 & -0.0561 & 0.0034 & -0.0001 & 0 & 0.0185 \\
-0.0014 & 0.0589 & 0.0680 & 0 & 0 & 0 & 0.0009 & 0.9943 & -0.0573 & 0.0012 & 0.0003 & -0.0003 \\
-0.0000 & -0.0184 & 0.0007 & 0 & 0 & 0 & 0 & 0.0013 & -0.0001 & 0 & 0 & 0.9998 \\
-0.0200 & 0.0355 & 0.9965 & 0 & 0 & 0 & 0.0064 & -0.0703 & 0.0001 & 0.0172 & 0.0037 & 0 \\
-0.0001 & 0.0000 & 0.0041 & 0 & 0 & 0 & -0.0214 & 0.0572 & 0.9981 & 0.0001 & 0.0000 & 0 \\
0.0001 & -0.0002 & -0.0064 & 0 & 0 & 0 & 0.9997 & 0.0008 & 0.0214 & -0.0001 & 0 & 0 \\
-0.9929 & -0.0007 & -0.0188 & 0 & 0 & 0 & 0 & 0 & 0 & -0.0842 & 0.0821 & 0 \\
-0.0017 & 0.0005 & 0.0146 & 0 & 0 & 0 & 0 & 0 & 0 & -0.6895 & -0.7242 & 0 \\
0.1173 & 0.0004 & 0.0122 & 0 & 0 & 0 & 0 & 0 & 0 & -0.7192 & 0.6847 & 0 \\
0 & 0 & 0 & 0 & 0 & -1 & 0 & 0 & 0 & 0 & 0 & 0 \\
0 & 0 & 0 & 0 & -1 & 0 & 0 & 0 & 0 & 0 & 0 & 0 \\
0 & 0 & 0 & -1 & 0 & 0 & 0 & 0 & 0 & 0 & 0 & 0
\end{array}\right]
$$

which leads to the following decomposition for the system:

$$
z_{co}^T = \left[z_4 \; z_5 \; z_6 \; z_7 \; z_8 \; z_9 \; -x_6 \; -x_5 \; -x_4 \right]; \qquad z_{c\bar{o}} = \begin{bmatrix} z_1 \\ z_2 \\ z_3 \end{bmatrix} \tag{11.20}
$$

with $z_i = \sum_{j=1}^{n} M_{ij} x_i, \quad i = 1, ..., n$. Finally, the vector zco may be decomposed into the available and unavailable parts as follows:

$$
z_{co}^T = \left[\xi_e \; \xi_m \right]; \quad \text{with} \quad \xi_e^T = \left[z_4 \; z_5 \; z_6 \; z_7 \; z_8 \; z_9 \right]; \quad \xi_m^T = \left[z_{10} \; z_{11} \; z_{12} \right]
$$

ξ_e is the vector of unmeasured variables, and ξ_m is the vector of available states. Matrices of the decomposed system are given by:

$$A_{co} = \begin{bmatrix} -38.85 & 28.99 & -0.00 & 0.17 & 0.01 & -0.01 & -5.10 & 0.04 & 1.00 \\ 2.33 & -50.29 & -0.26 & -0.24 & 2.72 & -2.0948 & -0.0205 & 0.0002 & 0.004 \\ 0.01 & -0.44 & -38.67 & -2.33 & -0.32 & -0.17 & 0.0327 & -0.0003 & -0.0064 \\ 0.01 & 0.07 & -28.69 & -29.23 & -0.04 & -1.26 & 2.25 & -0.1898 & 2.81 \\ -0.01 & 1.29 & -0.08 & 0.11 & -38.99 & 0.81 & -0.07 & 1.6664 & 1.60 \\ 0.04 & 0.28 & 7.71 & -1.26 & -0.51 & -39.77 & -0.31 & -1.5741 & 1.35 \\ 0 & 0 & 0 & -9.38 & -0.01 & 1.11 & -167.007 & -0.0190 & -47.90 \\ 0 & -0.00 & -0.00 & -0.24 & 21.29 & -20.38 & 9.6467 & -29.4080 & 11.05 \\ 0.00 & 0 & -0.00 & 0.25 & 20.27 & 21.41 & 12.23 & -0.0033 & -40.74 \end{bmatrix}$$

$$B_{co} = 10^3 \begin{bmatrix} 0.1040 & -0.6657 & 0 \\ -0.3872 & 0.2181 & 0 \\ 0.0052 & 1.2322 & 0 \\ 0.0252 & 0.0145 & 0 \\ 0.0057 & 0.1856 & 0 \\ 0.1403 & 0.0522 & 0 \\ 0.0332 & 0.0332 & -0.0076 \\ 0.0708 & 0.0708 & 0 \\ -0.0789 & -0.0789 & 0 \end{bmatrix}, \quad C_{co} = \begin{bmatrix} 0 & 0 & 0 & 0 & 0 & 0 & 0 & 0 & -1 \\ 0 & 0 & 0 & 0 & 0 & 0 & 0 & -1 & 0 \\ 0 & 0 & 0 & 0 & 0 & 0 & -1 & 0 & 0 \end{bmatrix}$$

with the following control constraints:

$$u_{max} = \begin{bmatrix} 9200 \\ 18446 \\ 200 \end{bmatrix}, \quad u_{min} = \begin{bmatrix} 2300 \\ 18446 \\ 100 \end{bmatrix} \tag{11.21}$$

The observer may be designed, at this stage, for the decomposed system. To this end, matrix T is chosen such that only the part $z(t) = T z_{co}(t)$ is estimated. Further, matrix P is chosen to ensure the asymptotic stability of matrix $D = T\, A_{co}\, P$. In fact, in this case, matrix T is given by:

$$T = \begin{bmatrix} 1\,0\,0\,0\,0\,0\,0\,0\,0 \\ 0\,1\,0\,0\,0\,0\,0\,0\,0 \\ 0\,0\,1\,0\,0\,0\,0\,0\,0 \\ 0\,0\,0\,1\,0\,0\,0\,0\,0 \\ 0\,0\,0\,0\,1\,0\,0\,0\,0 \\ 0\,0\,0\,0\,0\,1\,0\,0\,0 \end{bmatrix} \text{ and } \quad P = \begin{bmatrix} 1\,0\,0\,0\,0\,0 \\ 0\,1\,0\,0\,0\,0 \\ 0\,0\,1\,0\,0\,0 \\ 0\,0\,0\,1\,0\,0 \\ 0\,0\,0\,0\,1\,0 \\ 0\,0\,0\,0\,0\,1 \\ 0\,0\,0\,0\,0\,0 \\ 0\,0\,0\,0\,0\,0 \\ 0\,0\,0\,0\,0\,0 \end{bmatrix}$$

According to Eqs. (5.29), the matrices D, G, and E are computed.

$$D = \begin{bmatrix} -38.858 & 28.9925 & -0.007 & 0.1784 & 0.0119 & -0.0185 \\ 2.3306 & -50.2968 & -0.2666 & -0.2478 & 2.7231 & -2.0948 \\ 0.0147 & -0.4418 & -38.678 & -2.3376 & -0.3296 & -0.1756 \\ 0.0119 & 0.0735 & -28.6943 & -29.2308 & -0.0443 & -1.2648 \\ -0.0105 & 1.2947 & -0.0824 & 0.1128 & -38.9935 & 0.813 \\ 0.0442 & 0.2873 & 7.7115 & -1.263 & -0.5140 & -39.7722 \end{bmatrix}$$

$$G = 10^3 \begin{bmatrix} 0.1040 & -0.6657 & 0 \\ -0.3872 & 0.2181 & 0 \\ 0.0052 & 1.2322 & 0 \\ 0.0252 & 0.0145 & 0 \\ 0.0057 & 0.1856 & 0 \\ 0.1403 & 0.0522 & 0 \end{bmatrix}, \; E = \begin{bmatrix} -1.0078 & -0.0489 & 5.1034 \\ -0.0042 & -0.0002 & 0.0205 \\ 0.0064 & 0.0003 & -0.0327 \\ -2.8104 & 0.1898 & -2.2577 \\ -1.6047 & -1.6664 & 0.0718 \\ -1.3549 & 1.5741 & 0.3185 \end{bmatrix}$$

For the reconstruction error, one may choose the limits as follows:

$$\varepsilon_{max}^T = \begin{bmatrix} 1 \; 1 \; 0.5 \; 1 \; 1 \; 1 \end{bmatrix}; \quad \varepsilon_{min}^T = \begin{bmatrix} 0.5 \; 0.5 \; 0.25 \; 0.8250 \; 0.5 \; 0.5 \end{bmatrix}$$

For the matrix H, we choose to assign the following closed-loop eigenvalues $\{-170; -55; -51\}$, which leads to the following choice of matrix H:

$$H = \begin{bmatrix} -170 & 0 & 0 \\ 0 & -55 & 0 \\ 0 & 0 & -51 \end{bmatrix}$$

It is worth noting here that the remaining closed-loop eigenvalues are the $n - m$ stable ones coming from the open-loop system [14]. Hence, solving equation $F\,A + F\,B\,F \; = \; H\,F$ leads to:

$$F = \begin{bmatrix} 0.0010 & -0.0044 & 0.0319 & 0.0725 & -0.0791 & -0.0944 & 1.0349 & -0.0000 & 0.4162 \\ 0.0001 & -0.0005 & 0.0000 & 0.0001 & 0.0000 & -0.0002 & 0.0017 & -0.0000 & 0.0007 \\ 0.0148 & -0.0774 & 0.1421 & 0.0036 & -0.1591 & -0.2023 & -0.0352 & -0.0024 & 0.0864 \end{bmatrix}$$

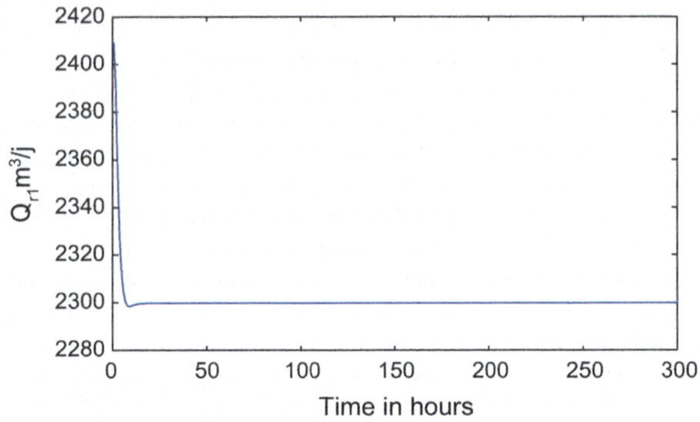

Fig. 11.7 Evolution of the internal recycled flow Qr1

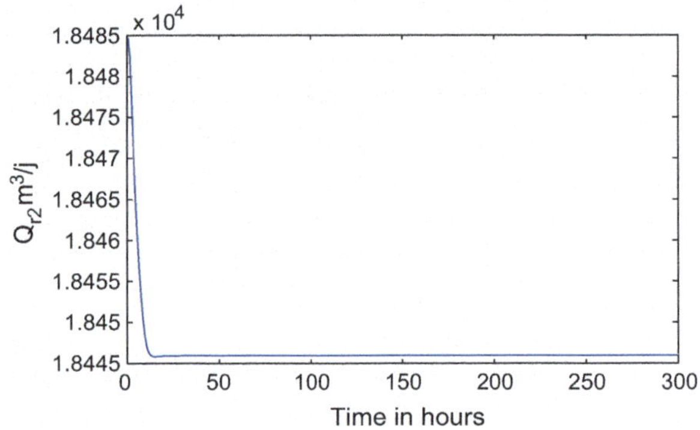

Fig. 11.8 Evolution of the internal recycled flow Qr2

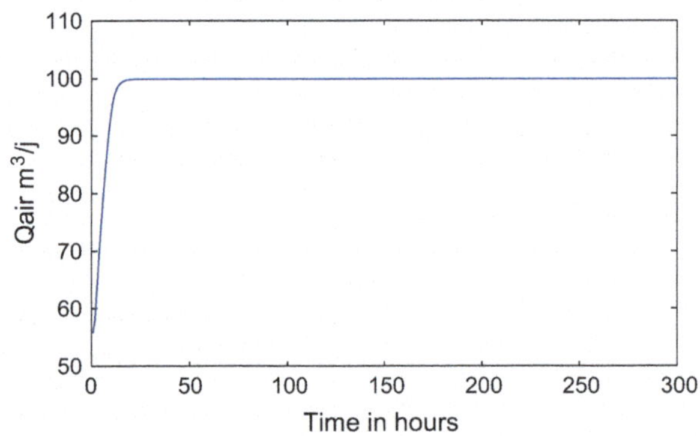

Fig. 11.9 Evolution of the dissolved oxygen Qair

Conditions of Theorem 5.1 are easily checked and are given by the vector

$$\tilde{M}_c\, q_\varepsilon = 10^4\,[-3.3981 \;\; -1.1 \;\; -0.1018 \;\; -0.0010 \;\; -0.0044 \;\; -0.0017 \;\; -0.0014 \\ -0.0037 \;\; -0.0034 \;\; -1.6974 \;\; -0.55 \;\; -0.0506 \;\; -0.0005 \;\; -0.0020 \\ -0.0016 \;\; -0.0008 \;\; -0.0018 \;\; -0.0014]^T < 0,$$

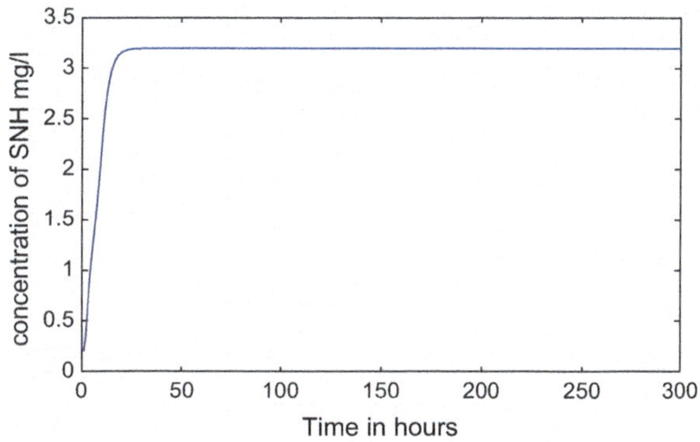

Fig. 11.10 Evolution of the ammonium S_{NH}

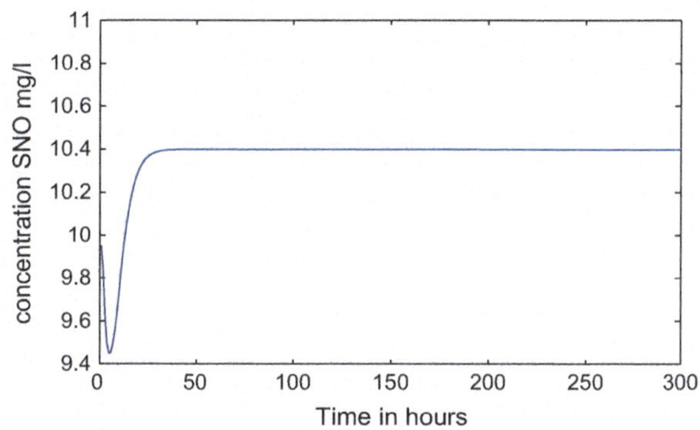

Fig. 11.11 Evolution of the nitrate S_{NO}

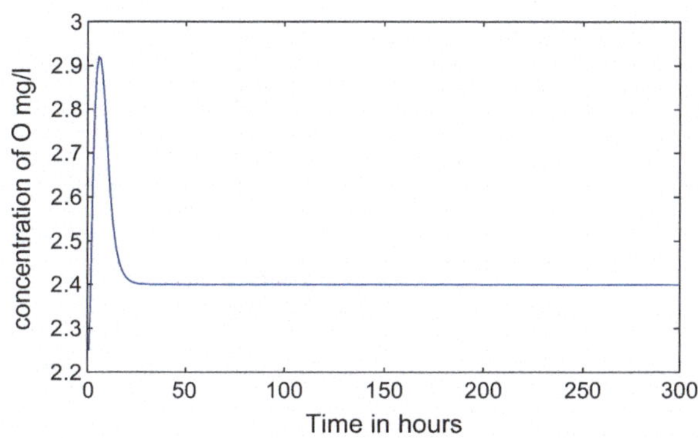

Fig. 11.12 Evolution of the oxygen O_2

which is a strictly negative vector. One may conclude that all required conditions are satisfied and hence, the observer-based controller as proposed is able to monitor the WWTP guaranteeing asymptotic stability and respect of all constraints for the control and the observation error.

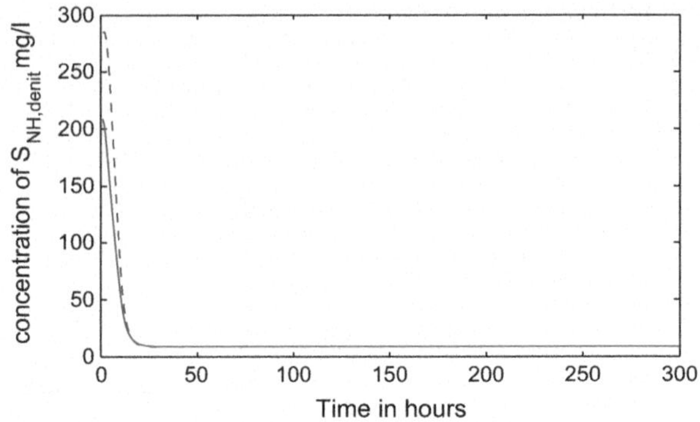

Fig. 11.13 The concentration of $S_{NH,denit}$ and its estimate

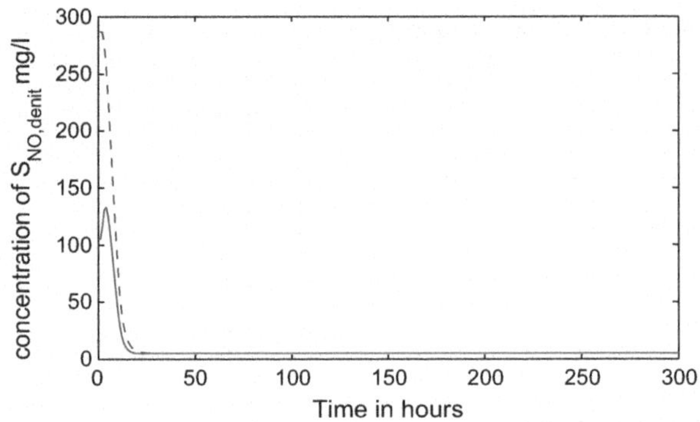

Fig. 11.14 The concentration of $S_{NO,denit}$ and its estimate

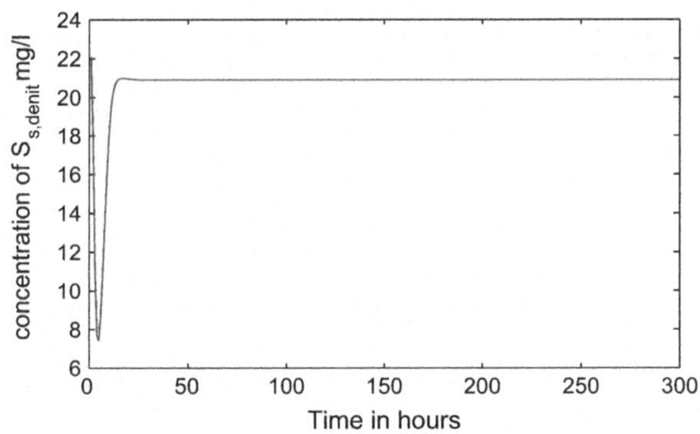

Fig. 11.15 The concentration of $S_{S,denit}$ and its estimate

11.3.2 Simulation Results

Figures below are devoted to present the evolution of all variables of the system. In fact, the observer-based controller, as defined in the section above, is applied to the WWTP. Estimated values are compared to the simulated ones from the nonlinear model. As general remarks, asymptotic stability is obtained, all constraints are

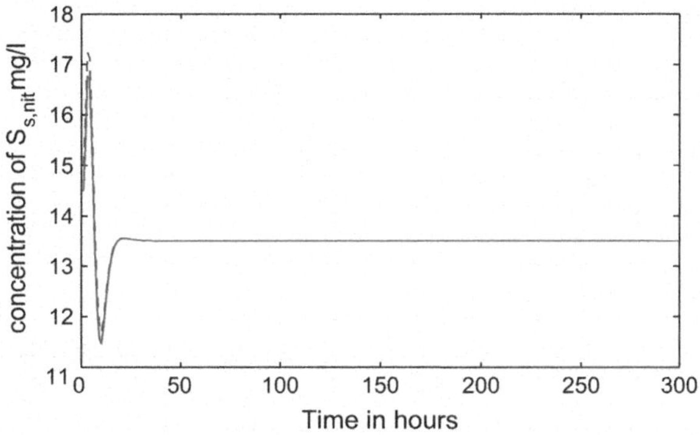

Fig. 11.16 The concentration of $S_{S,nit}$ and its estimate

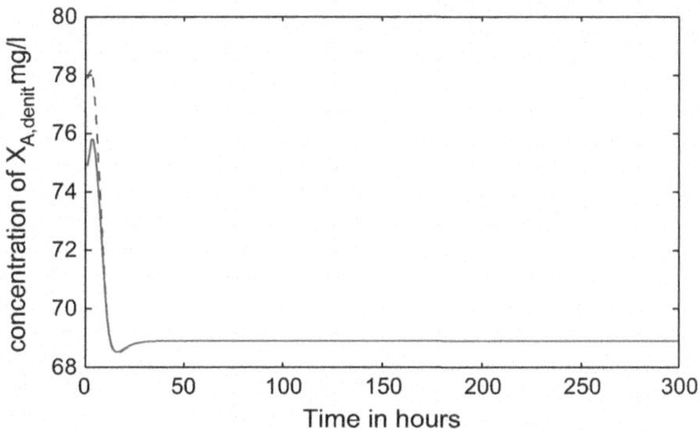

Fig. 11.17 The concentration of $X_{A,denit}$ and its estimate

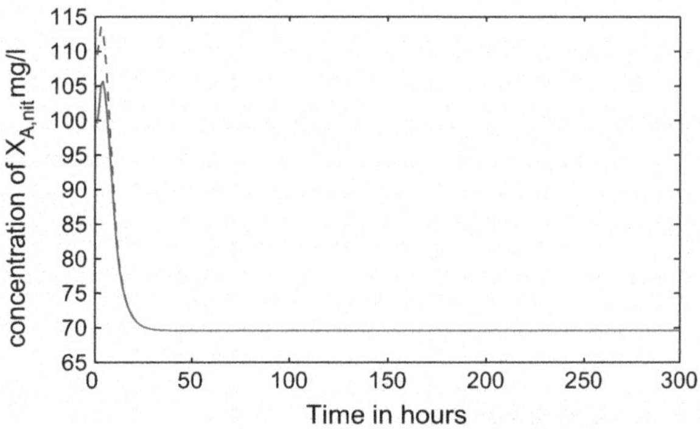

Fig. 11.18 The concentration of $X_{A,nit}$ and its estimate

respected and the amount of all non-desired organic matter is reduced in the output to the desired values. Further, the limits imposed to the estimation errors are also respected. First, the control evolution is presented in Figs. 11.7, 11.8, and 11.9. Respect of all given control constraints is clearly noticed. Second, Figs. 11.10, 11.11, and 11.12 show the output evolution from the initial values

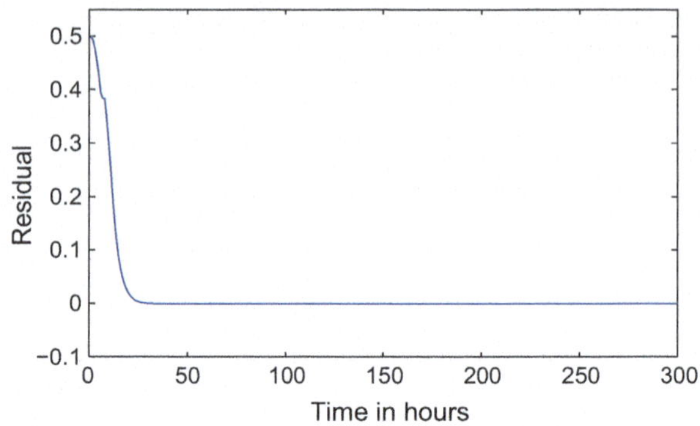

Fig. 11.19 Estimation error ($S_{S,nit}$)

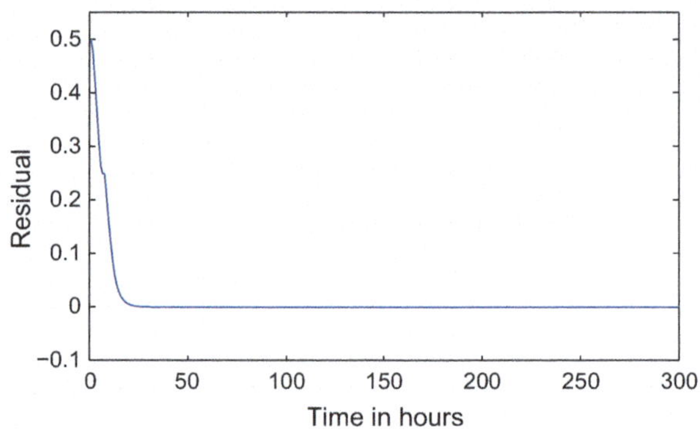

Fig. 11.20 Estimation error ($S_{S,denit}$)

$$y_o = Cx_o = C[100 \;\; 300 \;\; 15 \;\; 1 \;\; 10 \;\; 2.25 \;\; 74 \;\; 200 \;\; 22 \;\; 200 \;\; 100 \;\; 300]^T = [15 \;\; 1 \;\; 10]^T.$$

Finally, from Figs. 11.13, 11.14, 11.15, 11.16, 11.17, and 11.18 one may note that the convergence, for all estimated states, is obtained. Hence, and in practice, the concentrations of the organic matter are reduced and converge to the desired values. Furthermore, the Figs. 11.19 and 11.20 are devoted to show that the reconstruction error limits are really respected and this is clear from the figures.

11.4 Conclusion

In this chapter, the minimal order observer in the control loop of a nonlinear system with input constraints is introduced. In fact, observer, as software sensor, in the framework of positive invariance techniques is used to control the linearized model of a WWTP. For this process, linearization leads to some constraints on the control. Further, state variables are unavailable to measure and more than that no adequate sensor exists. Hence, the introduction of the observer is of great interest. The positive invariance techniques that had emerged as very efficient to handle similar problems of constrained control are successfully used to control the nitrogen removal process. The observer-based constrained control, as presented above may compete with approaches in easiness, applicability, and computing effort. In fact, all the needed computation is achieved offline and once the design finished, the control law is easy to implement on the process. Further, It is true that in the computational steps some trial and error tests are necessary; however, with the background available for the choice of the observer and the matrix H assigning the closed-loop poles [17], the computation effort is sensitively reduced. On the other hand, the evolution of the closed-loop system, as presented in the figures above, with the designed control law shows its efficiency and the success of the controller to the reduction of the organic waste. Other case study as spacecraft Rendezvous has been developed in the literature using gain scheduled control of linear systems subject to actuator saturation [29].

References

1. M. Ait Rami , H. Ayad, F. Mesquine, Enlarging ellipsoidal invariant sets for constrained linear systems. Int. J. Innov. Comput. Inf. Control **3**(5), 1097–1108 (2007)
2. G. Bastin, D. Dochain, On line estimation and adaptive control of bioreactors. Anal. Chim. Acta **243**, 324 (1990)
3. C.A. Belchiora, R.A. Araujo, J.A. Landeckb, Dissolved oxygen control of the activated sludge wastewater treatment process using stable adaptive fuzzy control. Comput. Chem. Eng. **37**, 152–162 (2012)
4. A. Benzaouia, The resolution of equation XA + XBX = HX and the pole assignment problem. IEEE Trans. Autom. Control **39**(10), 2091–2095 (1994)
5. A. Benzaouia, F. Tadeo, F. Mesquine, The regulator problem for linear systems with saturation on the control and its increments or rate: an LMI approach. IEEE Circuit Syst. I **53**, 2673–2680 (2006)
6. K.K. Biasizzo, I. Skrjanc, D. Matko, Fuzzy predictive control of highly nonlinear pH process. Comput. Chem. Eng. **21**, s613–s618 (1997)
7. B. Boulkroune, M. Darouach, M. Zasadzinski, S. Gille, D. Fiorelli, A nonlinear observer design for an activated sludge wastewater treatment process. J. Process Control **19**, 1558–1565 (2009)
8. D. Dochain, Design of adaptive controller for nonlinear stirred mnk bioreactors: extension to the MIMO situation. J. Process Control **1**, 41–48 (1991)
9. O. Galan, J.A. Romagnoli, A. Palazoglu, Robust H_∞ control of nonlinear plants based on multi-linear models: an application to a bench-scale pH neutralization reactor. Chem. Eng. Sci. **55**, 4435–4450 (2000)
10. T.F. Gustafsson, K.V. Waller, Nonlinear and adaptive control of pH. Ind. End. Chem. Res. **24**, 809–817 (1992)
11. M. Henze, C.P. Leslie Grady, W. Gujerm, G.V.R. Maraism, T. Matsuo, *Activated sludge Model No.1.* I.A.W.Q., Scientific and technical Report No. 1, (1987)
12. J.B. Lassere, Reachable controllable sets and stabilizing control of constrained systems. Automatica **29**, 531 (1993)
13. D.G. Luenberger, An introduction to observers. IEEE Trans. Autom. Control **AC-16**(6), 596–602 (1971)
14. F. Mesquine, *Contribution à la commande des systèmes dynamiques discrets avec contraintes sur les entrées par application du concept d'invariance positive.* Cadi Ayyad University, thèse de doctorat de troisième cycle (1992)
15. F. Mesquine, D. Mehdi, Constrained observer for linear continuous time systems. Int. J. Syst. Sci. **27**(12), 1363–1369 (1996)

16. F. Mesquine, *Contribution à la commande des systèmes linéaires à entrées contraintes par les observateurs et nouvelles méthodologie de placement de pôles*. Cadi Ayyad University, thèse de doctorat d'état (1997)
17. F. Mesquine, F. Tadeo, A. Benzaouia, Regulator problem for linear systems with constraints on control and its increment or rate. Automatica **40**(8), 1387–1395 (2004)
18. F. Mesquine, F. Tadeo, A. Benlamkadem, Constrained regulator problem for linear uncertain systems: control of a PH process. Math. Probl. Eng. (2006). doi:10.1155/MPE/2006/51874
19. F. Nejjari, A. Benhammou, B. Dahhou, G. Roux, Nonlinear multivariable control of a biological wastewater treatment process, in *4th European Control Conference* (Bruxelles, Belgique, 1997)
20. F. Nejjari, E. Dahhou, A. Benhammou, G. Roux, Nonlinear multivariable adaptive control of an activated sludge wastewater treatment process. Int. J. Adapt. Control Signal Process **13**, 347–365 (1999)
21. F. Nejjari, J. Quevedo, Predictive control of a nutrient removal biological plant, in *American Control Conference*, june 30–july 2 (Boston, Massachusetts, 2004)
22. M. O'Brien, J. Mack, B. Lennox, D. Lovett, A. Wall, Model predictive control of an activated sludge process: a case study. Control Eng. Pract. **19**, 54–61 (2011)
23. O. Pérez, F. Tadeo, P. Vega, Robust control of pH control plant, in *Proceedings of the IEEE Conference on Control Applications* (Albany, 1995)
24. Y. Smetsa Ilse, J.V. Haegebaerta, R. Carretteb, J.F. Van Impea, Linearization of the activated sludge model ASM1 for fast and reliable predictions. Water Res. **37**, 1831–1851 (2003)
25. A. Stare, N. Hvala, D. Vrecko, Modeling, Identification, and validation of models for predictive ammonia control in a wastewater treatment plant-a case study. ISA Trans. **45**(2), 159–174 (2006)
26. S.W. Sung, I.B. Lee, pH control using a simple set point change. Ind. Eng. Chem. Res. **34**, 1730–1734 (1995)
27. F. Tadeo, M. J. Grimble, Controller design using linear programming for systems with constraints. Part 1: Tutorial Introduction; Part 2: Controller Design; Part 3: Design Examples, IEE Comput. Control Eng. J. **12**, 273–276 (2002), **13**, 49–52, 89–93 (2003)
28. H. Zhao, S.H. Issacs, H. Soeberg, M. Kummel, Nonlinear optimal control of an allmating activated sludge process in a pilot plant. J. Process control **4**, 33–43 (1994)
29. B. Zhou, Q. Wang, Z. Lin, G. Duan, Gain scheduled Control of linear systems subject to actuator saturation with application to spacecraft rendezvous. IEEE Trans. Control Syst. Technol. **22**(5), 2013–2038 (2014)

General Conclusion

We were interested, in this book, to the general control problem with constrained inputs in both cases of continuous-time and discrete-time systems. That is the problem of limited inputs within symmetrical and nonsymmetrical constraints. It is the main link of all the presented results here. Further, constraints on the increment or rate of the control are also studied in this book. The rate or increment constraints are also non symmetric and are not nested in the control constraints as it is usually considered in the literature. Furthermore, the case of inputs and states constraints was also presented. The extension, when the system parameters are not perfectly known, has been obtained. Hence, robustness of such regulators is also under study in this book.

To handle these problems, two main approaches were used throughout this book. The first one is the exploitation of the so-called positive invariance concept. The method has been developed to avoid these constraints, maintaining the closed-loop system in a linear behavior region. Guaranteeing positive invariance of these regions leads to the validity when functioning of the linear behavior of the system once initialized inside the region. Hence, necessary and sufficient conditions of positive invariance of domains induced in the space by input constraints are established. We have then treated, respectively, the uncertain systems case, the rate or increment constraints case with and without disturbances, singular systems with PD control case, observer-based control for both regular and singular systems case. In all these works, the invariance conditions are linked to a pole assignment procedure popular for constrained systems obtained by the resolution of the algebraic equation $XA + XBX = HX$ for regular systems and $XA + XBXE = HXE$ for singular ones. Thence, closed-loop stability of constrained systems is deduced and design steps are precised for all cases.

At the end of this part, the LMI's tool to solve similar problems based on positive invariance techniques is introduced. Obtained conditions are reformulated using the Haar's lemma under matrix inequalities. It must be pointed out that performance specifications in terms of position of closed-loop eigenvalues can be easily incorporated in the design steps, which makes it a powerful technique.

A. Benzaouia et al., *Saturated Control of Linear Systems*, Studies in Systems, Decision and Control 124, https://doi.org/10.1007/978-3-319-65990-9

The second approach, in dealing with constraints, is based on letting the system to saturate using a technique of writing the saturating system as a convex combination of some linear systems. This idea was introduced recently in the literature. The application of this approach begins with the problem of static output feedback. Hence, sufficient conditions of asymptotic stability are dealt with. The results obtained with state feedback control are extended to the case of output feedback control. The second application was dedicated to solve the problem of handling nonsymmetrical constraints. In fact, contrary to positive invariance methods, this technique could not handle nonsymmetrical constraints. Hence, two methods were introduced to deal with asymmetrical constraints in the framework of LMIs. The first one needs an assumption that the system matrix must be non-singular. This may be seen as a conservative assumption. However, a second enhanced approach has been proposed to remove the restrictive assumption. More general result for asymmetrical constraints in framework of LMIs is obtained. Besides, these results extend those of the same authors developing unsaturating controllers working inside a region of linear behavior. Third application of the second approach to saturating systems was devoted to the introduction of the delay. Partition of the delay interval was introduced leading to less conservative results as pointed out in the literature. Effectively, some unfeasible problems become feasible. On the other hand, a delay-dependent criteria were worked out leading to stabilizability conditions for delay saturating systems. A comparison with previous works, but not in the saturating case, is achieved to show that the obtained conditions are less conservative.

As last application of the approach of convex writing of the saturating system, two-dimensional systems with constrained inputs and delays are presented. Sufficient conditions of asymptotic stability have been derived. The synthesis of the required controllers has been given under LMIs form.

It is worth here to recall that illustrative and academic examples were given, along all the previous chapters, to show the applicability and the effectiveness of the presented methods.

The last chapter was devoted to present the application of some of the previous results on some real processes. The first one is the control of a pH-process where in a tank the pH has to be controlled. Positive invariance techniques were applied in the context of considering the pH-process as an uncertain one. Simulation results emphasize the success of the obtained results.

For the second application, observer, as software sensor, in the framework of positive invariance techniques has been used to control the linearized model of a Waste Water Treatment Plant (WWTP). As usual, the linearization leads to some constraints on the control. Further, state variables are unavailable to measure and more than that no adequate sensor exists. Thence, the use of our technique was very adequate and successful.

Index

A. Benzaouia et al., *Saturated Control of Linear Systems*, Studies in Systems, Decision and Control 124, https://doi.org/10.1007/978-3-319-65990-9

The manufacturer's authorised representative in the EU is Springer Nature Customer Service Centre GmbH, Europaplatz 3, 69115 Heidelberg, Germany. If you have any concerns regarding our products, please contact ProductSafety@springernature.com

Printed and bound by CPI Group (UK) Ltd, Croydon, CR0 4YY

29/07/2026

02180578-0004